PLANT CHROMOSOMES

by

ÁSKELL LÖVE and DORIS LÖVE

with 29 figures in the text

1975 · J. CRAMER

In der A.R. Gantner Verlag Kommanditgesellschaft

FL-9490 VADUZ

Available in the USA from:
ISBS, Inc., 10300 S.W. Allen Boulevard,
Beaverton, Oregon 97005

Printed in Germany
by Strauss & Cramer GmbH, D-6901 Leutershausen
1. VIII. 1975
ISBN 3-7682-0966-0

This book is dedicated to

Cyril Dean Darlington,
Åke Gustafsson,
Albert Levan,
Arne Müntzing,
George Ledyard Stebbins.

PREFACE

The biota of the world are made up of cells which contain the substances concerned in heredity and, therefore, diversity. These elements, which we call chromosomes and genes, control the development of the individual and the evolution of life. Most of the hereditary material is transmitted from one generation to another with the chromosomes. Evolution depends, initially and fundamentally, on the changing structure and functional realization of the chromosomes and their genes, and on their transmission mechanisms and the manner by which the effects of the genes come into visible being and are preserved and altered through time. Therefore, understanding the chromosomes is significant for perceiving genetics and basic for the interpretation of evolution and its processes and, indeed, life itself and its history and origin.

Chromosome studies play their part in our comprehension of population structure. In the early years of genetics they gave us the concept of polyploidy and of rare changes in the linear arrangement of the chromosomes with all what that has come to mean for the understanding of the species, its origin, isolation, and further development. Such changes in the chromosomes, especially so-called fragmentation, are extremely rare, but it has been observed that they may increase considerably through certain artificial environmental influences and then even create conditions which will be adversary to the health and existence of the individual. These effects, even on human cells, can most adequately be observed through studies of the chromosomes of some plants. Since the observation of such phenomena is widely important, the study of the chromosomes has become necessary for teaching and research in all parts of the life sciences and in technology as well.

The chromosomes and their behavior can be studied in certain living cells with or without staining and then desirably by aid of phase or interference microscopes, although observation of preserved material with an ordinary light microscope is preferred for several reasons. In order to make them available for such a study, the dividing tissues have to be rapidly hardened or made insoluble, sometimes after pretreatment or a kind of narcosis, and then they must be selectively preserved without causing appreciable distortion of the elements of the nucleus. This is called fixation, and chromosome studies are usually performed after various processes of staining of fixed material in selected chemicals. The procedure is simple and opens a wide field of observation to those who master its techniques.

This text is intended as a kind of conducted tour in the elementary practical aspects of chromosome studies. We have tried to tell the reader about the phenomena involved, give him a quick explanation of their significance, and encourage him to try for himself by aid of the simple methods we describe. The book does not claim to explain everything, and it does not pretend to be exhaustive on what it does attempt. We do not intend to give a comprehensive survey of cell or chromosome behavior, although we hope the reader will become interested in looking for such information in more special textbooks. But we have placed a certain emphasis on technical details which frequently are left out of more elaborate manuals.

There are many to whom a practical introduction of this kind could appeal. Not only students in various botanical or biological fields, and then notably cytogenetics, plant breeding, taxonomy, and geobotany. But also those whose work calls for a general knowledge of the implication of this subject in the industrial world, where the future of unborn generations or the health of the individual himself may be affected by radiation from tools and machines or by drugs, food additives, or other chemicals that break or shatter chromosomes and create all kinds of mutations which may remain with us for generations.

It is evident that such a book cannot be compiled without consulting numerous other texts or without advice from many colleagues during the generation of our own work in this field. It will be obvious to the initiated that we have drawn substantially upon many of the current texts in cytogenetics in several languages, in addition to shorter reports too numerous to be mentioned or referred to. If we have inadvertently adopted or adapted where we should have sought permission, we hope it will be excused as oversight or ignorance. We want to acknowledge our debt to the authors of the works that we have consulted and to our colleagues from three continents for direct or indirect advice and encouragement and then particularly to the designer of modern chromosome study, Professor Cyril Dean Darlington of Oxford. We are also grateful to our students, likewise too numerous to be individually mentioned, for their patience in experimenting with various methods since our first preliminary and mimeographed recommendations in cytotechnology were made available to them, first at the University of Manitoba, then at the Université de Montréal, and later at the University of Colorado. Two of the latter must be mentioned, Professor Priyabrata Sarkar, now of the Department of Botany, University of Toronto, and Professor Brij M. Kapoor, now of the Department of Biology, Saint Mary's University in Halifax, because without their substantial help in abstracting the immense literature in the field of plant cytotechnology this book could not have been completed. The forerunner of the book was, however, a collection of a few mimeographed sheets with simple formulas of fixatives and methods, compiled, in the early 1940's, by our teachers, Professor Arne Müntzing and Professor Albert Levan, of the Institute of Genetics of the University of Lund. For their initial encouragement and advice we are for ever indebted.

Contents

INTRODUCTION

It is characteristic of the sciences, and indeed of all fields of knowledge, that a thorough understanding of the aggregate requires a detailed comprehension of its segregates. This is perhaps nowhere as evident as in modern physics, which owes its power to the discovery and understanding of the atom and its still smaller consituents of matter, leading to a coherent and comprehensive theory of matter and energy together. But it is also manifest in the life sciences, and then especially after the acceptance of the theory of evolution and the increased understanding of the cell and its nucleus with its chromosomes, genes and DNA. The chromosomes (fr. Greek, *chroma,* color, and *soma*, body, Waldeyer 1888) are threadlike structures that become visible during cell divisions. They are the site of or perhaps rather made up of the genes (fr. the Greek suffice *genos*, born, produced, Johannsen 1903; previously used in the combination *pangenes* by de Vries 1901, modified from Darwin's 1859 *pangenesis*), which are the smallest functional units of heredity and consist of DNA chains (deoxyribonucleic acid). The study of the chromosomes, genes and DNA has led to the explanation of the origin of life as given by molecular biologists and biochemists, whereas the processes of evolution have become elucidated through studies by cytogeneticists. These investigations have revealed that in every living being certain gene mutations (fr. Latin *mutos*, alter, de Vries 1901), genetic recombination and natural selection create and shape the diversity that is typical of the gene-pool, which is defined as the totality of the genes of an interbreeding population existing at a given time (Dobzhansky 1951). These studies have also shown that changes in the chromosomes, either structural and gradual or numerical and abrupt, are responsible for the creation and moulding of the reproductive isolation, which effectively separates different gene-pools. The biological species with its reproductive barrier has been identified with the gene-pool, a discovery which makes cytogenetics one of the most important tools of the modern taxonomist and geobotanist. This finding is also essential for workers in the fields of applied botany, especially in plant breeding, because it explains the limits of possibilities to improve the cultivated plants and furthermore points to new methods for their advancement.

Cytology (fr. Greek *kytos*, hollow place, cell) is the study of the cell and its elements, but its subdivision, karyology (fr. Greek *karyon*, nut, nucleus), investigates the nucleus (fr. Greek, kernel, nut, Brown 1833) and its constituents, the chromosomes, in rest and division. Chromosome phenomena are many and complex, but those of importance for the problems anticipated by botanists for whom this text is compiled are connected with mitosis (fr. Greek *mitos*, loop, and the general suffix *-is*, Flemming 1882) and meiosis (fr. Greek *meion*, smaller and *-is*, Farmer and Moore 1905, the term borrowed from rhetoric) in general, but particularly with variations in the number and morphology of the chromosomes and their behavior during the formation of the reproductive nucleus. Therefore, our descriptions of the chromosomes are restricted to these phenomena, whereas the reader will find more detailed and advanced information in some of the texts listed in a special bibliography in Appendix II to this book.

Cytotechnology or chromosome technology is the art of making chromosomes available for microscopic study: it is actually the branch of microtechnique which concerns the chromosomes. Although its method has alsways been basic for cytological investigations, it is only during the past three or so decades that others than cytologists have felt an urgent need to employ its technique to various other fields. This has led to the necessity of rationalizing and simplifying the methods employed. The need for a comprehensive text on the basic phenomena of chromosome botany and the technique by which these organelles are made visible has become increasingly evident to students working with plant taxonomy and other fields of evolutionary botany, but also to plant breeders and agricultural research workers and others concerned with various kinds of environmental pollution. It is the aim of the authors to try to fill this need of an introductory text for those, who have had no previous contact with these fields, and perhaps also to whet their appetite for more detailed studies by aid of the advanced texts referred to in the bibliography.

PART I: CYTOLOGICAL BACKGROUND

CHAPTER 1: THE SPOROPHYTE GENERATION

A. **Mitosis - the dividing cell**

a. **Preliminary**

All higher organisms that reproduce by sexual means are divided into two generations, the gametophyte (Hofmeister 1851) which produces the sexual generation or gametes (fr. Greek *gametes*, married couple, Strasburger 1871), and the sporophyte (Hofmeister 1851) which in the higher plants is the asexual generation and bears the gametophyte; the sporophyte is produced by so-called fertilization, when the male and female gametes unite to form the so-called zygote (fr. Greek *zygon*, yoke, Bateson 1902; Bateson and Saunders 1902). This initial sporophytic cell develops by aid of an equiproductive division, resulting in the formation of two new cells, which in turn divide, so that the process is repeated almost endlessly to form the mature individual and all its organs. Since all these cells arose from one cell, they include the same hereditary potentialities and the same complement of chromosomes. This is secured by a process called mitosis, in which the cell divides in a manner similar in all higher organisms to replicate the chromosomes exactly in every detail (Fig. 1). Mitosis is actually defined as a process by aid of which the chromosomes become split longitudinally into two equivalent parts that separate to two opposite poles of the cell to form two identical nuclei for the two new cells. It is a dynamic process so that one stage merges into the next without any definite line of demarcation. However, it has proven useful to part it into phases (Strasburger 1884; Heidenhain 1894) and attach to them names which provide a means of defining the part of the process under observation or discussion. The descriptions below are based on plant material, but although the chromosome movements are slightly different in some details in animals and fungi, they are generally the same in all eukaryotic organisms, which are the higher plants, animals and fungi (Margulis 1970).

The linkage group or DNA thread of bacteria and their viruses, which is devoid of basic proteins and has no mitotic apparatus and no nuclear membrane, has sometimes been named genophore or nucleoid and was

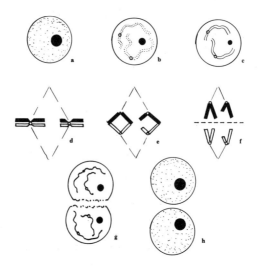

Fig. 1: Diagram of mitosis. a. Interphase. b, c. Prophase. d. Metaphase. e, f. Anaphase. g. Telophase. h. Two new interphase nuclei.

called chromoneme by Whitehouse (1965) (from Greek *chromo,* color, and *nema,* thread). Since a very similar term, or chromonema, has been used by Wilson (1896) and later authors for either the chromosome thread itself or for half-chromatids, or for the finest visible longitudinal subdivision of a chromosome, it might seem to be less confusing to call the prokaryotic thread chromolene (fr. Greek *linon,* thread, cf. Woods 1944, 1966), the spelling being slightly adjusted for pronunciation. The chromolene is a structure analogous but at least not directly homologous to the chromosomes of eukaryotic biota, and it often does not stain with the Feulgen technique. Prokaryotic cells seem to divide directly, mostly by so-called binary fission.

b. Interphase

The cell contains a clear, usually viscous, fluid in which there are numerous granular and rod-like bodies. This is the cytoplasm, which is kept in place by the cell wall, and surrounds the nucleus with its own thin membrane.

Most nuclei are not in a dividing stage. They are then recognizable in the microscope as spherical and lightly stained bodies without a definable inner structure. Although it is now known that this is the period of genetic or metabolic activity of the nucleus and the cytoplasm, the terms interphase (Lundegårdh 1912) or resting stage are generally used to describe the condition between the cell divisions, but difference ought to be made between the active interphase of dividing cells and the resting stage of mitotically inactive cells. Within each of the nuclei the chromosomes lie in the form of long threads which are more or less loosely coiled and pressed against the membrane that surrounds the nucleus. However, at this stage the chromosomes do not stain intensely with standard dyes, presumably because of the distribution of nucleoprotein. Sometimes small regions of the chromosomes remain visible during the interphase; they are known as chromocenters (Tischler 1920). Within each nucleus is one or more dense and rounded bodies, which are called nucleoli (Bowman 1840); they frequently need to be stained by special methods. There is evidence that the nucleoli are formed close to small regions of the chromosomes which are called nucleolar organizers (McClintock 1934). It has been suggested that the nucleolus contains a reserve of diverse substances which are utilized by the cell at various times, probably mainly during mitosis.

Cytophotometric, autoradiographic, and cytochemical studies have shown that the active interphase can be divided into three periods, for which Howard and Pelc (1953) suggested the designations synthetic stage (S), which is the period of DNA synthesis and chromosome replication, gap 1 (G_1) or the presynthetic initial growth period between the end of telophase and S, and gap 2 (G_2) which is the postsynthetic period of continued growth between S and prophase. According to Mazia (1974), the G_1 stage takes about eight hours, the S stage takes about six hours, G_2 requires about five hours, whereas mitosis and cell division need about one hour for completion under ideal conditions.

c. **Prophase**

When an interphase cell has reached a certain stage of development, it starts to divide. The exact nature of this beginning is unknown, although it is understood that a synthesis of proteins, nucleic acids and other substances must take place first. At the time when the mitotic division begins, it goes into so-called prophase (fr. Greek *pro*, before, and phase, Strasburger 1884, 1905), when the chromosomes start to contract and thicken into a coiled mass which becomes stainable and recognizable as definite forms. The substance in nuclei and chromosomes which absorbs the stain was named chromatin by Flemming (1880); it is made up of acidic compounds of DNA with proteins. Each chromosome at this stage is long and thin and divided along its length into two strands or identical halves called chromatids (McClung 1900), which are often twisted around each other. The chromosomes slowly contract, and at least a part of the contraction appears to be associated with the development of a new coil, at right angles to the old, which grows from many small to relatively few larger rings so that the chromosome contracts to about one-fifth of its original length. There is one small region of each chromosome, called the centromere (fr. Greek *centron,* center, and *meros*, part, Waldeyer 1903), which is undivided at this stage and holds the chromatids together. The nucleoli disappear late in prophase, and the end of the stage is indicated by the disruption of the nuclear membrane so the contents of the nucleus can mingle with the cytoplasm.

d. **Metaphase**

Following the sudden breaking up of the nuclear membrane and the dissolution of the nucleolus, the chromosomes continue to contract and move towards a central position in the cell where they arrange themselves in the equatorial portion of the spindle apparatus to form the so-called metaphase (fr. Greek *meta*, between, after, and phase, Strasburger 1884,1905). The spindle is a bipolar collection of fibers which is produced in late prophase as a result of an interaction between a clear zone around the prophase nucleus, the nuclear sap, and the centromeres (Bajer 1957, 1961). The spindle stains very weakly and appears as a clear and structureless fibrous area going out from each end of the cell and reaching its greatest width in the equatorial plane. It is made up of continuous fibers running from pole to pole and chromosomal fibers which become attached to the centromeres, which are a part of a non-staining constriction on each chromosome, always occurring at a position which is constant for and characteristic of each particular chromosome. The centromere, which is also called kinetochore (fr. Greek *kinesis*, movement, and *chorein*, to move forward, J.A. Moore, in Sharp 1934) or primary constriction (Darlington 1937), does not become coiled, and it remains unchanged through prophase and metaphase. Some chromosomes have, in addition, one or more secondary constrictions (Darlington 1926), frequently close to one of the ends so that a small headlike piece, or satellite (Navashin 1912), is formed. These latter constrictions are not concerned with chromosome movements but frequently mark the positions of the nucleolar organizers mentioned in connection with the interphase nucleus, since it is at these sites that the spherical nucleoli are formed early in the mitotic cycle. The constancy in position of both primary and secondary constrictions is an important characteristics for the recognizing of individual chromosomes. The centromeres in plants tend to line up more or less uniformly in the central region of the spindle, with the chromosome arms lying at random, whereas in animals and fungi they incline to form a circle at the periphery.

During metaphase the outline of the chromosomes becomes clearly defined and they reach their maximum density and staining ability. It is at this stage that their number can most easily be counted, because they are in the same plane and usually less intertwined than earlier or later in the division. Frequently they can be seen to be longitudinally separated into chromatids. It is also evident at this stage, that although they vary in length and in the relative position of the centromere, the chromosomes are, in a diploid species, usually met with in pairs the identical members of which are called homologues.

e. **Anaphase**

The next stage in the mitotic cycle is the anaphase (fr. Greek *ana*, back, and phase, Strasburger 1884, 1905), when the centromeres, which until now have been the only undivided parts of the chromosomes, themselves divide all at once so that each chromatid gets a centromere of its own. With the chromosome arms trailing, the centromeres move towards opposite poles along the spindle, repelling one another or pulled by some unknown mechanism. At a later stage of the anaphase the spindle itself lengthens and becomes narrower at the equator and forms two closely packed groups of chromosomes at each of the poles.

f. Telophase

On arrival at the poles, the chromosomes slowly lose their staining properties, as if the prophase was being reversed. They uncoil to become long and thin threads, and nucleoli begin to form as does a new nuclear membrane. The chromocenters, if met with at all, soon become evident. These changes, or telophase (fr. Greek *telos*, the end, and phase, Heidenhain 1894), occur at exactly the same time in both daughter nuclei and bring the two new cells into a new interphase stage by the process of cytokinesis (fr. Greek *kytos*, cell, and *kinesis*, movement, Whitmann 1887), when a cell plate or phragmoplast (fr. Greek *phragma*, fence, and *plastos*, formed, Errera 1888) appears at the equatorial region of the spindle to form the cell wall. The two cells are identical in every detail and the replication of the chromosomes has been accompanied by an exact reproduction of all the genes, or DNA chains, which they carry.

It must be observed that this description of mitosis is a generalized one and based on what is typical of most flowering plants, which are characterized among other by having distinctly unicentric chromosomes, or chromosomes with a single centromere. There are several deviations, and in the families Cyperaceae and Juncaceae and at least some algae and insects the centromere is not single but the chromosome is either polycentric or holocentric or with a lateral or diffuse centromere. This affects the chromosome movements at anaphase in certain ways, as will be described later, and makes it possible for the chromosomes to break in various places to increase the chromosome number without increasing the chromosome volume or the number of genes. Breakage of that kind is called agmatoploidy (fr. Greek *agmos*, a fracture, breakage, Malheiros-Gardé and Gardé 1950).

B. Basic number and polyploidy

The mitotic process is the same irrespective of the number of chromosomes, which in the somatic or body cells of the organism is said to be diploid (fr. Greek *diploos*, double, and *id*, a combination proposed by Strasburger 1905; *id* is Weismann's (1885) term for hypothecial structural units of the nucleus, derived from Nägeli's 1884 idioplasm, fr. Greek *idios*, peculiar). Whereas the sex cell after reduction division or meiosis is said to be haploid (fr. Greek *haploos*, single, and *id*). Diploid is always designated as $2n$, haploid as n. Frequently, species of a genus are characterized by a multiple of the same low number of gametic, or sex cell, chromosome sets or haplomes (Heilbronn and Kosswig 1938, 1966) in which each individual chromosome of a set is represented only once; a corresponding genetical term which is frequently confused with this cytological one is genome (Winkler 1920) which designs the sum total of the genes in which each allele occurs only once. Allele (Johannsen 1909) is an abbreviation of the term allelomorph (fr. Greek *allelon*, one another, and *morphe*, form, Bateson and Saunders 1902) for genetic characters that are alternative of one another. Such a series of multiple haplomes (Fig. 2), which may reach up to 100 or more times the original haplome,

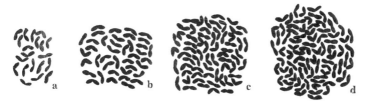

Fig. 2: A polyploid series in *Polygonum*, *2x, 4x, 6x, 8x*. From Löve and Löve (1956).

as in some ferns, is called a polyploid series (fr. Greek *polys*, many, and *ploid*, an arbitrary term formed by analogy with haploid and diploid, Winkler 1916), and its constituents are called polyploids, with the Greek prefixes mono, di, tri, tetra, penta, hepta, octo, etc. depending upon the multiples they represent. At any level, however, haploid specifies the chromosome number of the gametic cells or the haplophase, and diploid that of the somatic cells or the diplophase (Renner 1916), irrespective of the degree of polyploidy of the plant, although diploid is also the term for the lowest diplophase number in the polyploid series. The lowest haploid number of a polyploid group, including only a single haplome, is said to be monoploid (Langlet 1927a), and it is also the basic number of the series, designated as x. The other members of a polyploid series are marked as $2x$, $3x$, $4x$, $6x$, $8x$, etc., whereas the only correct utilization of the n-series above that of $2n$ is when marking the triploid or polyploid nature of the polar nuclei of the embryo-sac, because then it is compared with the real haploid number of the reduced egg cell and its diploid number after fertilization.

Basic numbers may be as low as $x = 1$ or 2, but most frequently they are $x = 3, 4, 5, 6, 7,$ or 11. Higher numbers are supposed to be formed by combination of the lower, as, e.g., $6 + 7 = 13$, or by so-called secondary polyploidy (Darlington and Moffett 1930) which is a term denoting polyploids with a chromosome complement in which particular chromosomes in the basic set are represented more frequently than others. Some groups of plants are characterized by very high basic numbers which certainly are derived and secondary in origin.

Polyploids are of different kinds, depending upon their origin (Fig. 3). First, they are either autoploids or alloploids (Kihara and Ono 1926; Clausen, Keck and Hiesey 1945), the former having originated through the duplication of essentially identical haplomes of the same species, the latter by the duplication of the chromosome number of a more or less sterile hybrid of two more or less distinct species. Second, it has been found practical to subdivide each of these kinds into two groups. Then, typical or strict autoploids are named panautoploids (fr. Greek *pan*, all, Löve and Löve 1949), if they have been formed by duplication of the haplomes of a single originally diploid population, preferably of strictly autogamous, or self fertilizing, plants or at least of reasonably purebred clusters of allogamous, or cross-fertilizing, species. Such a polyploid is supposed to be very similar to the original local population or deme (fr. Greek *demos*, a people, cf. Langlet 1971; the term was originally proposed as a suffix by Gilmour and Gregor 1939). But it is isolated from the parent deme reproductively and is likely to have different adaptive properties caused by the difference in chromosome number alone. Polyploids of this kind may occur in most average sized populations in the very low frequency of 2-5 per thousand, but they rarely seem to survive more than a single generation. The alternative type of pattern is

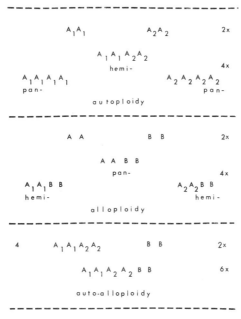

Fig. 3: Kinds of polyploidy.

that of the rare and difficult hybridization of two species which are so widely distinct that their haplomes and chromosomes are almost completely non-homologous so that the hybrid is completely sterile. Its polyploid, however, will immediately become a constant alloploid, or panalloploid, extremely successful and combining the morphological and physiological characteristics of the parent species which had already been condoned and stabilized by natural selection, and with essentially the properties of fully fertile diploids, thus the classification of such tetraploids as amphidiploids (fr. Greek *amphi*, on both sides, and diploid, Navashin 1927). In between these two extremes, without any distinct limit, are the hemialloploids (fr. Greek *hemi*, half, Löve and Löve 1949), which are formed from only partially sterile hybrids between species which differ in segmental arrangements of their chromosomes (segmental polyploids, Stebbins 1947), and the hemiautoploids, which are produced either from more or less fertile intraspecific hybrids or by differentiation

13

of the chromosome set of successful panautoploids or through hybridization of related panautoploids. These two intermediate groups constitute the majority of known natural cases and seem also to enclose a considerable number of cultivated polyploids. The chromosomes of autoploids tend to differentiate so that multivalent formation at meiosis is reduced and bivalent formation increased; that is called bivalization, which is not to be confused with diploidization, or the doubling of the chromosomes in haploid cells or hyphae of fungi. When in complex polyploids the alloploid populations have derived from demes at different levels of ploidy. e.g. a diploid AA and an autoploid BBBB so the alloploid contains a different number of haplomes from each parent, the term autoalloploid is appropriately used (Kostoff 1939), whereas a complex polyploid formed from diploid and variously polyploid demes based on numerous original haplomes has been called compiloploid (fr. Greek *compilo*, pillage, scrape together, de Wet and Harlan 1966).

In this connection it needs to be emphasized that the making of a polyploid from any kind of stabilized gene-pools always results in the formation of a strong barrier to gene exchange that prevents or decreases miscibility through hybridization, since hybrids between the new polyploid and its diploid or lower polyploid parent gene-pool, if at all produced, are inevitably sterile so that following generations either fail to appear or show a distinct loss of vitality. The causes of this will be discussed later, in connection with the description of meiosis, but it should be noted here, that such a reproductive barrier is the basic requirement of the species category when it is defined biologically. Since polyploidization is an instantaneous process that takes one or rarely two generations, it has been called abrupt speciation, in contrast to the slower and progressive process of the linear differentiation of the chromosomes, which leads to gradual speciation that will be considered later (Valentine 1949; Löve 1954, 1964).

Polyploidy invariable affects quantitative characters as the size of the cells and the form and size of the organs of the plant, although for the latter there seems to be no general rule. Polyploids have larger pollen than diploids of the same polyploid series. Their stomata are also significantly larger, when a polyploid derived from a certain diploid is compared with the parent strain or species, although the fact that the size of stomata and certain other cells may be affected by selection even between local populations, or demes, requires that such comparisons be made with caution. Some polyploids, especially autoploids, frequently demonstrate a giant growth, and often the leaves of autoploids are broader and shorter than those of the diploids, and also generally darker because of increased thickness.

Successful natural polyploids, even autoploids, also differ in qualitative characters which are caused by divergence at the genetic level, through the establishment of new genes in a manner similar to that increasing the diversity of any gene-pool, as will be discussed in some detail on a later page. That, however, is apparently a process even slower than bivalization of autoploids, as shown among other by the fact that many polyploids which must be very old as indicated by their distribution still closely resemble their diploid relatives in most morphological and physiological features.

Polyploidy has been of considerable importance for the evolution of biota, as shown by the fact that perhaps 25-30% of all higher plants are polyploid, and the significance of this condition for the genesis of many cultivated plants is evident since most of these species are characterized by having high chromosome numbers. Their success as cultivated plants is matched by their apparent survival ability under severe conditions in nature, as is mirrored for instance in the increase in their frequency with latitude and altitude. It was long believed that this was caused by a direct influence of extreme environments on the formation of polyploids, because experiments showed that temperature shocks at a certain stage after fertilization greatly increase the number of polyploid seeds. More recently this has been found to be a fallacy, since many more polyploids are apparently formed in the tropical and temperate regions than in the coldest areas of the earth, although their frequency is much lower within these largest of floras (20-25%) than in the small floras of the cold regions (85-90%) (Fig. 4), but that difference is the result of severe selection (Löve and Löve 1971, 1974). It has been shown that polyploidy directly increases the genetical variation and, thus, creates a strong mechanism of a kind of preadaptation, which is the possession of the necessary adaptional gene variations which enable the plant to adapt to conditions to which it has not been previously exposed. Through this preadaptation the polyploids become clearly superior to their diploid relatives when forced either into higher altitudes by rising mountain chains, or into the short and cold summers of the northlands by continental drift of the lands on which they were formed.

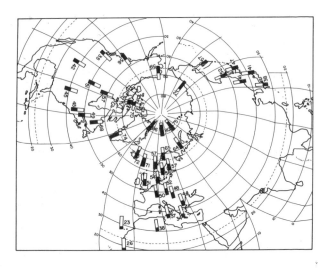

Fig. 4: Frequency of polyploids in non-alpine regions of the northern hemisphere. From Löve and Löve (1974).

A phenomenon that rarely may affect sex mother cells and only then lead to sudden polyploidy, is the so-called endomitosis (Geitler 1939) which is the replication of the chromosomes without a division of the nucleus. It results in endopolyploidy (Rieger and Michaelis 1958) - which has also been called didiploidy (de Litardière 1925) or polysomaty (Langlet 1927b) - the visible outcome of which may be occasional.polyploid cells or sectors in the tissues where it occurred. It is a rather common incident in cells at the upper limit of divisions in the meristem in roots of some families, especially in Aizoaceae and Chenopodiaceae, and may be the cause of reports of polyploid chromosome numbers in otherwise distinctly diploid species. Its evolutionary significance remains obscure, although it is apparently typical of certain specialized tissues, notably the endosperm or the tapetum cells of the anthers where repeated endomitotic divisions may lead to a high degree of polyploidy. Repeated endomitosis without subsequent division of the centromere and separation of the chromatids is also the cause of so-called multistranded or polytene chromosomes

(Koltzoff 1934), which are especially typical of the salivary glands of larvae of some dipterous insects, but also of the embryo suspensor of at least some plants in which such giant cells occur in the distal portion.

It is of interest to note that after treatment with very low concentrations of colchicine, various derivatives of naphthalene, and certain other poisons, a kind of endomitosis has been observed. This division, called c-mitosis (Levan 1938), is caused by that the poison prevents the formation of the spindle and, therefore, the separation of the divided chromosomes. The procedure has been found to be very useful in the production of artificial polyploids of cultivated plants and trees (Eigsti and Dustin 1957). A similar treatment with these chemicals occasionally results in a reduction in the chromosome number of polyploids, a phenomenon which some authors have inappropriately called somatic reduction or diploidization, terms that have a different meaning as originally defined. This lowering of the chromosome number after such treatments still has not been satisfactorily explained.

In certain mitotic tissues, it is sometimes observed that a spindle is formed with three or four poles. The chromosomes are distributed randomly to these poles so that cells with deviating chromosome numbers are produced. Such divisions are said to be multipolar. They have been observed most frequently in galls and in cancerous tissues and are especially frequent after chloralization or treatment with alcohol, ether, acetone, and many esters. When dividing cells are treated with some natural alkaloids, as colchicine and podophyllin, or with barbiturates, some sleeping drugs with chloralhydrate, some sulphonamides and insecticides or herbicides, the formation of the spindle is frequently disturbed or completely put out of function.

Among serious aberrations in chromosome division is also the so-called stickiness, or a difficulty or inability of chromosome separation because some chromosomes stick together at anaphase. In some cases this seems to be inherited, but it is frequently caused by irradiation, or by certain chemicals.

C. Aneuploidy and supernumeraries

Normally individuals or populations of a species possess chromosome numbers that are exact multiples of the basic number for the group. They are then said to be euploid (fr. the Greek prefix *eu-*, good, true, Täckholm 1922). Loss or addition of individual chromosomes is frequently caused by so-called non-disjunction of a chromosome pair at the meiotic anaphase I after an apparently perfect metaphase pairing, but such a numerical change is also known to occur after mitotic disturbances; this inevitably results in irregular multiples, a condition which is called aneuploidy (fr. the Greek negative prefix *an-*, and euploid). Such a condition, for several chromosomes, also occurs after selfing or back-crossing of sterile triploids, pentaploids or hybrids the meiotic pairing of which is normally upset. Each chromosome pair of an euploid series is said to be disomic in the diploid, tetrasomic in the tetraploid, or polysomic in the polyploid (Blakeslee 1921, 1922). In aneuploid plants, monosomics occur when one chromosome of a pair is missing, nullisomics when an entire pair has been lost, and trisomics when an additional chromosome of a pair is met with; a more detailed terminology including more complex cases has recently been proposed by Khush (1973). Monosomics are an important tool for establishing the place of certain genes on individual chromosomes; for this purpose they have been used extensively for chromosome mapping in *Triticum*, whereas trisomics have been used for similar purposes in *Datura*.

Stabilized aneuploidy which characterizes related taxa which because of it become distinguished by having slightly different basic numbers, is called dysploidy (fr. the Greek prefix *dys-*, un- or mis-, Tischler 1937). It is frequently accompanied by some structural changes in the chromosomes concerned and seems to be an important feature in the origin and evolution of new groups at the generic level.

When supernumerary chromosomes are the result of simple aneuploidy, they are normal chromosomes the additional occurrence of which is occasional and disturbing for the individual and its development because of increased or decreased genetical material. In some plants, however, peculiar chromosomes additional to the ordinary complement are occasionally met with. These chromosomes have a limited genetical effect and differ frequently from normal, or A, chromosomes, in being smaller and sometimes perhaps heterochromatic (fr. Greek *heteros*, different, Heitz

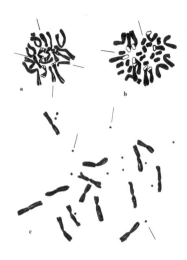

Fig. 5: Metaphase plates with B-chromosomes as indicated by arrows. a. *Secale cereale,* 2n = 14 + 4B. b. *Festuca pratensis,* 2n = 14 + 15B. c. *Allium cernuum,* 2n = 14 + 13B. a. from Müntzing (1944). b. after Bosemark in Müntzing (1971). c. after Vosa in John and Lewis (1965).

1928), i.e. their chromatin may not follow the normal degree of contraction and staining of ordinary chromatin, or euchromatin (fr. Greek *eu-,* true), and they seem to carry few genes. These are the so-called B-chromosomes (Randolph 1928; cf. Müntzing 1959), and they do not pair with A-chromosomes and at least frequently do not form chiasmata at meiosis. Some observers have attributed a limited adaptive importance to the possession of B-chromosomes, whereas others found no indication for this. When they occur in high numbers, they seem to affect the pairing of the A-chromosomes by reducing their frequency of chiasmata and thus cause some decrease in the fertility of the individual. The number of B-chromosomes may vary from cell to cell and from individual to individual (Fig. 5), although there seem to be some B-chromosomes which are mitotically stable so that all cells will have exactly the same number of them, as may be the case in all the species of the genus *Rhinanthus* so far studied.

A phenomenon more common and perhaps more significant than the occurrence of aneuploidy or B-chromosomes, is the presence of so-called fragments, which are parts of chromosomes which lack a centromere. They have a short life expectancy because of their limited mobility and they are soon eliminated during mitosis. Fragments are the result of a breakage of a chromosome either at meiosis or at mitosis; their genetical effect is mainly deleterious, because when they are eliminated important genetical material is lost.

Fragmentation at mitosis is sometimes caused by some accident during the division, but more frequently it is the result of some environmental influence. Although occasional fragments are perhaps of a limited significance, their number seems to increase with the age of the individual, or perhaps they tend to accumulate with the increased seniority of the mitotic line despite a high frequency of elimination. Of still more concern is the observation that cancer is characterized by fragmented chromosomes, although fragmentation is not necessarily its primary cause. It has been demonstrated that fragmentation may be the result of ionizing radiation, or of ultraviolet light, or of radiation from television apparatus or other appliances in daily use, or it can be caused by heat, artificial colors or carbon compounds, some food additives or spices or even coffee, many drugs and various other chemicals, and by certain virus diseases, which, thus, are said to be carcinogenous or mutagenic, because before their concentration reaches the chromosome breaking stage, they will have caused invisible but serious damage to the genes or DNA chains, so-called loss mutations.

Chromosomal changes and gene mutations which occur in cells that remain in the mitotic cycle may affect the health and well-being of the individual, promote its aging through various unexplained physiological disturbances, or even cause its premature death. But if they occur in cells which later develop into sex cells, then these changes or loss of genetical material may affect later generations infinitely or until the damaged chromosomes are eliminated.

The chromosome breaking effects of irradiation and chemicals are most easily established through studies of fragmentation in the root-tips of diploid plants with large chromosomes, as, e.g., *Allium* and *Tradescantia*. For such studies, the plant is grown in a field of irradiation or in media into which the chemicals have been mixed, and the cells are checked for the occurrence of irregularities after one or a few cell

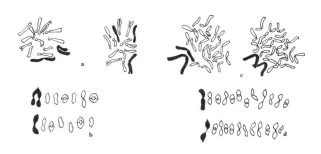

Fig. 6: Sex chromosomes. a. *Acetosa pratensis*, mitosis; to the right male with $2n = 12 + X + 2Y$, to the left female with $2n = 12 + 2X$. b. *Acetosa pratensis* meiosis; upper row $6_{II} + XYY$, lower row $6_{II} + XX$. c. *Melandrium dioicum* spp. *dioicum*, mit sis; to the left male with $2n = 22 + XY$, to the right female with $2n = 22 + XX$. d. *Melar drium dioicum* spp. *rubrum*, meiosis; upper row $11_{II} + XY$, lower row $11_{II} + XX$. a. from Ōno (1935). b. after Kihara and Yamamoto (1932). c. from Westergaard (1940). d. from D. Löve (1944).

divisions. On basis of such studies it is possible to establish safe limits for mutagenic or carcinogenous agents that must be used in industry or in other human endeavours. By aid of careful investigations it has been shown, that there is no lower limit for the chromosome breaking effects of radiations, so distance and screening are most important when radiation has to be allowed, and X-raying ought to be limited only to absolutely unavoidable cases in hospitals. Even some chemicals, which in low concentrations seem to be essential to life, as, e.g., ordinary table salt, may become mutagenic or carcinogenous when these concentrations are surpassed even slightly.

D. Sex chromosomes

In dioecious plants and many animals, the determination of sex is frequently connected with a pair of chromosomes, two similar ones in the so-called homogametic sex, and by one of these and one or two other chromosomes which pair partially with it in the heterogametic sex. These are called allosomes or sex chromosomes (Wilson 1906), as contrasted to the normal autosomes (Montgomery 1904) or the somatic complement of chromosomes not directly affecting sex determination. In the homogametic sex, both the chromosomes are designated as X, whereas the deviating member or members of the pair in the heterogametic sex are called Y; the latter is sometimes absent. These chromosomes were originally discovered as an uneven or heteromorphic bivalent in the meiosis of insects and mosses but were much later found also in some seed plants and other higher organisms (Fig. 6). As long as their role was obscure, they were called accessory chromosomes (McClung 1900), a term frequently misused for B-chromosomes in recent decades. Later studies have shown that heterochromosomes connected with sex are most easily and safely identified in good slides of the mitotic metaphase, at least in diploid plants.

Dioecious plants and animals with X and Y chromosomes may have the same cytological mechanism of sex determination, and the cytologically heterozygous sex is usually the male, except in birds and butterflies. However, the genetical mechanism that actually decides about the sex of the individual is generally of two fundamentally different kinds. In animals and mosses without a Y chromosome in the heterogametic sex ($XX : XO$), and also in many taxa with an $XX : XY$ cytological mechanism, sex is apparently determined by a kind of a balance between the number of X chromosomes and the number of autosome sets, so that a diploid individual with two haploid autosome complements ($2A$) and two X chromosomes will form the homogenous sex and be a female, whereas an individual with $2A$ and only a single X will become a male. In such cases, the Y chromosome has been found to be genetically almost inert and without any sex determining genes. Since all individuals with an intermediate proportion of autosome sets and X chromosomes will become intersexual or hermaphrodite, this mechanism seems to counteract the development of polyploid dioecious species (Muller 1925). However, in probably the majority of both plants and animals, including all the mammals, the sex of the individual is mainly determined by the presence or absence of a Y chromosome, which in these taxa is packed with strong male determining genes, whereas the sex determining effect of the X chromosome is minor or nil and the autosomes mainly female determining. In that case polyploids may remain dioecious at least in sofar as the increase of the female determinants in the multiplied autosomes does not upset the strength of the male determinants in the Y. In some dioecious plants the male determinants in the Y have been found to be strong enough to dominate at least 12 autosome complements of a highly polyploid and purely dioecious taxon (Westergaard 1958; Löve 1969)

In some dioecious plants and animals, cells from certain tissues can be specially stained to reveal differences in nuclear structure connected with sex. In the dioecious species of the genus *Acetosa*, the interphase nuclei of the female plant are of a granular structure, whereas those of the male plant also have a deeply stained body, which makes it possible to distinguish between the sexes at any stage of development by aid of a simple cytological study. This body, which is called sex chromatin, was first discovered in female mammals, which have a largely heterochromatic X chromosome, whereas the genus *Acetosa* is characterized by having two mainly heterochromatic Y chromosomes. It has been suggested that the sex chromatin may be derived by condensation of heterochromatin of the second X in the mammals and the second Y in *Acetosa*, but other explanations may be equally plausible (Pazourková 1964; Mittwoch 1967).

E. Karyotype - chromosome form and size

In general, corresponding chromosomes of a pair are identical in size and form in different cells of the same individual and in members of the same deme and even of the same species, whereas different pairs may vary considerably in length and form within the same cell. Variations in size and form of the chromosomes between genera and related species, and sometimes between populations of the same species, are of importance in classification and for certain evolutionary considerations (Satô 1962); such variations are said to represent different karyotypes (Delone 1922; Jackson 1971). The karyotype of a taxon is frequently expressed by the drawing of a so-called karyogram (Chiarugi 1933; idiogram, Navashin 1921), in which the chromosome pairs are sometimes arranged after their size and the situation of the centromere either from camera lucida drawings or from enlarged microphotographs from which the individual chromosomes are cut out and pasted on a cardboard paper for reproduction. Karyograms are also often drawn diagrammatically based on measurements of the chromosomes, although all such measurements are difficult and inexact so that mainly the relative size and shape of the chromosomes can be correctly represented (Fig. 7). Some methods for the study of the karyotype are described on a later page.

The chromosomes of the same taxon usually extend gradually from small to long; sometimes they range in the order of ten-to-one, although usually

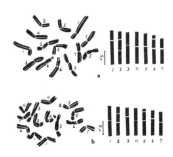

Fig. 7: Karyograms. a. *Aegilops comosa*. b. *Aegilops comosa* ssp. *heldreichii*. From Chennaveeraiah (1960).

their range is closer to one-to-two and frequently even less. The smallest chromosomes of plants are known from sedges, in which they look dot-like and are about 0.2 micrometer long, or at the limit of visibility in the light microscope; whereas in some Liliales, Magnoliales, and the Gymnosperms they may be more than 25 micrometers long and 2.5 micrometers thick (Fig. 8). The size of the chromosomes is usually similar in related species, but it may vary considerably within the same family, as, e.g., in Droseraceae, in which the relation between the volume of the chromosomes in some *Drosera* and *Drosophyllum* is as 1 : 1000. The chromosome size is apparently determined by the coiling of the chromosomes, since after special treatment affecting the coiling all the chromosomes in some or many of the cells of the tissue treated may remain uniformly smaller for a division or two. Such variations may, however, also be caused by special genes, but then the difference is persistent.

In some plant genera the chromosome size is distinctly bimodal, with a group of large and thick and small and thin chromosomes in a distinct mixture. In the genus *Rhinanthus* of the Scrophulariaceae all the species have fourteen medium sized and eight very small chromosomes (Fig. 9), and similar observations have been made in some mosses. There is a possibility that the small chromosomes in these taxa could be constant B-chromosomes, although conclusive evidence for that theory is lacking.

Fig. 8: Variation of chromosome size in plants. a. *Bolboschoenus maritimus*, 2n = 104. b. *Spergula marina*, 2n = 36. c. *Cucubalus baccifer*, 2n = 24. d. *Bellevalia romana*, 2n = 8. e. *Haemanthus katharinae*, 2n = 18. a. - d. after Blackburn (1933). e. after Satô (1959). All same magnification.

Fig. 9: Chromosomes of *Rhinanthus minor*, 2n = 14 + 8, of which the smaller may be permanent and constant B-chromosomes. After v. Witsch (1950). × 1000.

Fig. 10: Positions of the centromere. From left to right: median, or v-chromosome, submedian, or j-chromosome; subterminal, or i-chromosome; terminal.

When chromosome size is significantly different in groups of species classified in the same genus by authors of flora manuals, there is a valid reason to suspect that these taxa may be less closely related than indicated by some of their morphological characteristics, so they might be better separated in distinct genera.

The individual chromosomes may be recognized by aid of their morphology and then especially by the position of the centromere and by the localization of secondary constrictions, when these occur, or by aid of possible heterochromatic segments.

The position of the centromere can be classified as being median, submedian, or subterminal (Wilson 1928; Tischler 1951; Levan, Fredga and Sandberg 1964), resulting in what also are called v-, j-, or i-chromosomes (Kihara and Yamamoto 1932), whereas truly terminal centromeres are rare (Fig. 10). In another terminology (White 1940, 1945), these chromosomes are said to be metacentric, submetacentric, subtelocentric, acrocentric, or telocentric. Sometimes a secondary constriction may demarcate a short part of the chromosome, either intercalary or, most frequently, terminally. Such a terminal piece is called a satellite (Navashin 1912) and the chromosome itself a SAT-chromosome (from "sine acide thymonucleinico", or "without DNA", Heitz 1931), because of the supposed heterochromatic nature of the secondary constriction. A special case is the holocentric or polycentric chromosomes (Darlington 1937; Rieger, Michaelis and Green 1968), which are met with in agmatoploid biota; in the terminology by Wilson (1928), such centromeres were termed lateral.

26

As mentioned earlier, B-chromosomes may consist mainly of hetero-chromatin, in which the condensation and the staining cycle are out of phase with the normal euchromatin. It seems also to be the material of which sex chromatin is made. Heterochromatin is met with in certain parts of the ordinary chromosomes of plants and may then stain less than other parts and make the chromosomes look banded or cross-striped, especially after treatment with cold (Darlington and La Cour 1940), quinacrine mustard (cf. Caspersson and Zech 1973), or after using a Giemsa staining technique (Schweizer 1973; Schweizer and Marks 1974).

Studies of karyotypes require exactness in technique at every level, but when the material is suitable for such investigations, they can be rewarding and render information of considerable evolutionary and taxonomical significance.

CHAPTER 2: THE GAMETOPHYTE GENERATION

A. Meiosis - the reduction division

a. The male gametophyte

The higher plants form a life cycle of two generations of cells which are situated on the same individual, a haploid (n) gametophyte and a diploid ($2n$) sporophyte (Hofmeister 1851). The differentiation between these generations is through the meiotic division, during which the diploid chromosome number is reduced to the haploid one.

Meiosis is, with other words, the transition from sporophytic to gametophytic tissue. It consists of two successive nuclear divisions, accompanied by only one functional division of the chromosomes. It results in the formation of four nuclei, each of which has only half the chromosome number of the nucleus of the sporophyte (Fig. 11). In plants, meiosis is most easily studied in the pollen mother cells, and it is commonly so synchronized that all the cells of an anther, or even those of all the anthers in a flower, may be at the same stage of division at the same time.

Before the onset of meiosis there is an interphase, during which the nuclei enlarge considerably, and late in this phase the content of DNA increases much. It is not known what causes the difference between the mitotic and meiotic divisions, but the latter deviates mainly from the former by its long and complicated prophase.

I. The first meiotic division

1. Prophase I.

The first prophase is the stage of contraction, characterized by the reappearance of the chromosomes as stainable bodies. Since this is a longer and much more elaborate step than the prophase of the mitosis, its subdivision into some stages based on the appearance of the chromosomes is warranted.

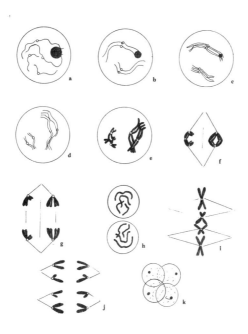

Fig. 11: Diagram of meiosis. a - e. Prophase I. a. Leptotene. b. Zygotene. c. Pachytene.
d. Diplotene. e. Diakinesis. f. Metaphase I. g. Anaphase I. h. Telophase I. i. Metaphase II.
j. Anaphase II. k. Interphase.

Leptotene (fr. Greek *leptos,* thin, and *taenia,* ribbon, v. Winiwarter
1900): At the beginning of meiosis, the large nucleus is transformed into
long and thin threadlike chromosomes, which are too long to be
individually distinguishable and show little of the relic coiling of the
somatic prophase. The chromosomes appear to be single and not divided
longitudinally. In good preparations of some material it is possible to see
a series of darker stained beadlike granules, called chromomeres (fr.
Greek *chroma*, color, and *meros*, a part, Wilson 1896; Fig. 12); they are
irregularly distributed along the length of the lighter staining thread.
In plants with large chromosomes, the chromomeres are of characteristic
sizes and have definite positions so it has been suggested that they may

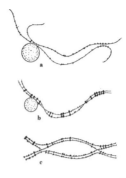

Fig. 12: Chromomeres and pairing in prophase I. a. Zygotene. b. Pachytene. c. Diplotene.

mark the situation of definite genes or groups of genes. During leptotene and throughout the meiotic first prophase, the nucleic membrane is intact and the nucleoli are distinct and attached to the nuclear organizers. The chromosome number is the diploid one, with one set which originally came from the male parent of the plant and another from the female parent.

Zygotene (fr. Greek *zygon*, yoke, and *taenia*, Gregoire 1907): When the homologous chromosome threads begin to move towards pairing, or synapsis, the next stage begins. The pairing usually starts at one end and proceeds along the whole length of the chromosomes successively, but it may also begin at both ends at the same time, or at more points along the chromosomes, but always in the same way in the same species. While they are pairing, the chromosomes twist around each other and contract. The pairing is a very exact process with chromomeres of similar size and at corresponding positions getting together. The synapsis may be complete or partial, depending on the exactness of the homology of the chromosomes, and it may also be affected by nutrition, temperature, specific genes, and variation in the linear arrangement of the chromosomes themselves. While this is taking place, the chromosomes become thicker and shorter. When the pairing is complete, the pair of homologous chromosomes is so close together that the nucleus appears to contain only the haploid numbe s; these pairs are from now on called bivalents (fr. Greek *bi-*, two, and *valens*, to be strong, active, Haecker 1892).

Pachytene (fr. Greek *pachys*, thick, and *taenia*, v. Winiwarter 1900): The chromosomes continue to contract and the paired chromosomes of each bivalent come to be still more closely associated and often become twisted around one another. It is inferred from observations of later stages, that during pachytene, in one or more places in each bivalent, exchanges of segments between homologous chromatids are taking place, in the so-called chiasmata (fr. Greek *chiasma*, two lines placed crosswise, Janssens 1909), although is not visible until at the next stage. The chiasmata are assumed to form the physical basis for crossing-over (Morgan and Cattell 1912), or to make it possible for the two chromatids to exchange parts and change or shuffle the linear composition of their genetical linkage groups (Fig. 12b).

Diplotene (fr. Greek *diploos,* double, and *taenia*, v. Winiwarter 1900): At the beginning of this stage, the chromosomes appear to lose their mutual attraction so they no longer are in close contact along their entire length. The bivalents tend to clump in the center of the nucleus, making observations difficult, but it is evident that in one or more places in each bivalent there are chiasmata which hold them together. In material with large and not too many chromosomes it can be seen that each chromosome is divided along its entire length except at the centromere, so that each bivalent clearly consists of two pairs of chromatids. The chromosomes are held together at the chiasmata where a union of homologous regions has been formed. During diplotene the chiasmata tend to move towards the end of the bivalent and become terminalized (Darlington 1929). The chromosomes then continue to contract and begin to repulse each other from the centromere onwards (Fig. 12c).

It needs to be mentioned here that the number of chiasmata per bivalent is characteristic for each taxon and connected with chromosome size and the position of the centromere. Small chromosomes tend to have few chiasmata which become fully terminalized, whereas long chromosomes may have numerous chiasmata of which some remain interstitial until at the end of the prophase and sometimes longer. Also, chromosomes with a median or submedian centromere frequently form chiasmata in both arms which give rise to a closed or ring-bivalent, whereas those with subterminal centromere usually form chiasmata only in the longer arm resulting in a rod-bivalent.

Diakinesis (fr. Greek *dia-*, apart, and *kinesis*, movement, Haecker 1897): When contraction of the chromosomes is near a maximum and the chromosome pairs have become well spread as round bodies with distinctive shape due to the terminalization of chiasmata, the last stage of the prophase has been reached. By the end of this stage, the major coil of the chromosome is usually completed and the nucleolus and also the nuclear membrane disappear.

2. Metaphase I

The first metaphase of meiosis begins when the dissolution of the nuclear membrane is complete. A spindle is formed and the bivalents move slowly to line up at the equatorial plane. Since each bivalent possesses two centromeres, however, they cannot lie in the same plane but instead homologous centromeres become oriented towards opposite poles on each side of the equator and at an equal distance from it. The arrangement is completely at random. The chromosomes stain deeply. This stage is especially suitable for studies of the frequency of chiasmata, and for observations of deviations from the normal condition as will be described on a later page.

3. Anaphase I

Homologous centromeres begin two move towards opposite poles. Each centromere remains undivided and attached to both the chromatids of a chromosome, contrary to mitosis when chromatids separate. These chromatids are homologous in some regions and sisters in others, i.e., they derive either from homologous chromosomes or from the same chromosome, depending on the previous occurrence of chiasmata and crossing-over. The chromatids are gradually loosened and the chiasmata slip apart, and the original chromosome complement is reduced to half when the chromosomes reach the poles.

4. Telophase I

When the chromosomes regroup at the poles of the cell, a stage similar to that of the mitotic telophase is reached. The chromosomes become uncoiled and form two nuclei. This stage may vary in length even within the same individual, and in some plants, as, e.g., *Trillium*, the anaphase of the first meiotic division goes straight into the prophase of the second division. Usually, a membrane is formed around the nuclei.

In many monocotyledons a wall is laid down between the nuclei to produce a two-cell or dyad stage; that is called successive pollen formation since the final wall is formed later. In most dicotyledons, however, the telophase is short, the chromosomes go through the second division without uncoiling in between, and all the cell walls are formed at the same time after the end of the second division, when a tetrad is formed; that is called simultane pollen formation.

5. Interphase

This stage is absent in many plants and short in others, although it is of a relatively long duration in plants with successive pollen formation. The main differences between meiotic interphase and that of mitosis are, that whereas the daughter chromosomes which form the interphase nucleus at mitosis are single threads, the half-bivalents which form the meiotic interphase are clearly double threads which are still held together by their undivided centromeres. And there is no further division of the chromosomes during this meiotic stage, which is only a kind of a strengthening prelude to the second and final meiotic division.

II. The second meiotic division

1. Prophase II

The second division serves to separate the genetically different chromatids of each original chromosome pair. If a distinct interphase has been entered, and a wall formed between the two nuclei of the dyad, then it is possible and indeed likely that the two nuclei will not go into the second division simultaneously. If there is no interphase, however, and the chromosomes have remained visible, the first telophase may pass imperceptibly into the second prophase. It is then a very short stage during which the chromatids contract and thicken and become more distinct and are visible as a pair of threads held together by the undivided centromere. A spindle appears in each part of the dyad, but the relative orientation of the two spindles varies considerably, although they are frequently at right angles to each other.

2. Metaphase II

At the end of prophase, the centromeres, which still hold the two chromatids together, move towards the equatorial plane of the spindle and form the metaphase plate. This is a stage when the chromosomes can easily be counted, because they are short and thick. Since their number is only the haploid one, many cytologists still prefer this stage for chromosome number determinations, especially when the number is high. However, several plates must be counted, particularly if meiotic irregularities have been observed at earlier stages or can be expected to have happened, because such disturbances may cause certain variations in the number of chromosomes that become evident at this stage and later.

3. Anaphase II

The centromeres divide longitudinally and move towards the poles, pulling with them the individual chromosomes or former chromatids.

4. Telophase II and interphase

Now we have the four groups of chromosomes which uncoil and extend and slowly lose their ability to stain. They gradually become invisible and form the new interphase to complete the making of the tetrad, which is the collective name of the four cells which later will form the pollen grains. Each of the tetrad cells has a haploid (n) set of the chromosomes of the original and mitotic pollen mother cell.

III. Pollen mitosis and pollen grains

The four potential pollen grains of the tetrad grow considerably in size after the meiotic division has ended, and eventually develop a thick wall. In order to form the male gametophyte, the pollen nucleus has to divide to form a generative and a vegetative nucleus before its development is completed. Of these, the generative nucleus divides again to form two sperm nuclei; this last division occurs before maturity of the pollen grain in about one-third of all genera of higher plants to form a trinucleate pollen grain, whereas in the majority of plants the pollen grain is binucleate and the division of the generative nucleus occurs in the pollen tube after germination (Fig. 13). The two pollen nuclei differ in appearance, the generative nucleus being compact and the vegetative nucleus diffuse and indistinct. Although pollen grain mitosis is an attractive material for studies of chromosome number and karyotype because of its haploid number, it seems to happen very fast and at a limited time so that locating anthers and pollen tubes containing this stage is often a very tedious, though rewarding, work.

The ripe pollen grain is, with exception of some water plants, as *Ceratophyllum, Najas*, and *Zostera*, usually covered with two membranes or walls, the intine and the exine. The latter is characteristically patterned and has a number of pores which make it possible to identify the genus and sometimes even the species from their pollen grains alone. This is done by palynologists, some of whom study the climate of bygone millennia by aid of such identification of pollen grains of plants which formed the changing vegetation of the past.

Most tetrads break into individual pollen grains at maturity, although some remain together or form larger groups. Durable tetrads or com-

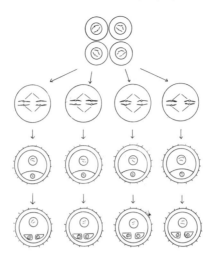

Fig. 13: Development of male gametophytes. Based on Müntzing (1971).

pound pollen grains are typical of the Ericales, and also of *Typha latifolia* as contrasted to other species of that genus. The pollen grains of Mimosaceae and Orchidaceae form clumps of numerous grains which are connected within the same intine and exine into so-called massulae (fr. Latin *massula*, a little clump) that in some orchids are again united into the pollinium, which includes all the pollen grains of a single theca of the anther. It ought also to be mentioned that in the Cyperaceae no tetrads are formed, but three of the four pollen microspores degenerate within the single pollen grain that results from each meiotic division.

IV. Asynapsis, desynapsis and cytomixis

Sometimes the pairing of the chromosomes does not materialize at prophase and metaphase of the first meiotic division so the entire process results in the forming of a single nucleus with the unreduced chromosome number. If the failure of pairing is complete already at early

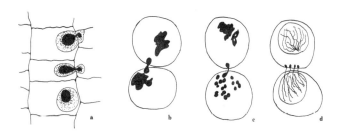

Fig. 14: Cytomixis. a. In mitotic cells. b, c, d. In pollen mother cells. a. After Tarkowska (1973). b, c, d. after Á. Löve (1943a).

prophase, this is called asynapsis (Beadle and McClintock 1928; Beadle 1930), but when the pairing appears to be normal at and before pachytene but fails at diplotene because no chiasmata were formed, the phenomenon is termed desynapsis (Sharp 1934). Since it is frequently difficult to ascertain if there actually was a pairing at all in early prophase, many cytologists prefer to include both processes under the former and then collective term. Asynapsis is usually caused by genes affecting chiasma frequency (Lin and Paddock 1973), or even by sex-linked genes (Á. Löve 1943b), but may sometimes be the result of environmental disturbances. For such cells there is actually no reduction division, and the resulting cell will be diploid with the chromosome number $2n$.

In agamosperms with a meiotic division, synapsis may seem to be normal during prophase, but then something breaks down and causes the formation of so-called restitution nuclei which are diploid and contain all the chromosomes of the mitotic pollen mother cell.

Rarely, chromatin from one cell is observed as passing into the cytoplasm of another, usually between pollen mother cells or between cells at early stages of meiosis, but also in mitotic divisions (Fig. 14). This is called cytomixis (Gates 1911). It is most likely an artifact caused by some environmental or mechanical disturbances perhaps even during fixation (Tarkowska 1973).

37

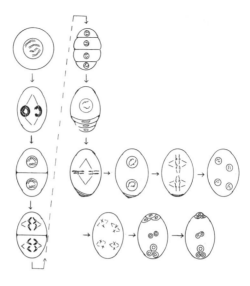

Fig. 15: Development of female gametophytes. Based on Müntzing (1971).

b. The female gametophyte

As we have mentioned, the male cells of the pollen grains or the microspores of higher plants are formed in large quantities within the anthers. Because of their being readily accessible and numerous, it is a relatively simple task to observe the meiotic divisions of the pollen mother cells that lead to the male gametophyte. The study of the development of the female gametophyte, however, is a more difficult task, because the female cells, or embryo-sacs (Hofmeister 1849), occur singly and embedded in the tissue of a young ovule and develop from one embryo-sac mother cell. This cell enlarges and elongates considerably prior to the meiotic divisions so that it becomes much larger than a pollen mother cell at a similar stage. However, meiosis in the embryo-sac mother cell is similar to that in the pollen mother cell and results also in four haploid nuclei when it has gone through all the same phases of the first and second meiotic division. The four haploid nuclei constitute the primary four-nucleate stage which is comparable to the tetrad stage of the pollen. They are usually arranged in a linear order. The way in which the embryo-sac continues its development varies considerably in different groups of plants. Although one, two, or even all four of the haploid nuclei may be concerned in the formation of the mature embryo-sac, which ultimately can consist of from four to sixteen nuclei, usually three of the primary nuclei degenerate and only the fourth, or the lowest one in the row, enlarges and goes through three successive mitotic divisions. After the first mitosis, one of the two new nuclei moves to the chalazal or proximal end of the sac, and the other to the micropylar or distal end (the micropyle is the opening between the integuments, or the envelopes of the ovule, through which the pollen tube will enter). Subsequently, two additional mitoses result in four nuclei at each end of the embryo-sac (Fig. 15).

One nucleus from each of the groups moves to the center of the embryo-sac where they eventually combine to form a diploid fusion or polar nucleus or the secondary embryo-sac nucleus. Cell walls are formed around each of the three nuclei remaining at the ends of the sac. Those at the proximal end form the so-called antipodals, whereas the three at the distal end form the egg apparatus, which consists of one egg cell and two synergids. Deviations from this typical or primary embryo-sac structure are characteristic of different families; they have frequently been employed in classification of plants into orders and are of importance for grouping of these orders into still higher categories.

B. Meiosis in structural hybrids

The description of the meiotic process given above is based upon the assumption that the plant is a diploid individual with chromosome pairs which are made up of essentially identical and homologous chromosomes. Such pairs form normal bivalents in the first prophase and metaphase which then separate at the first anaphase to produce new cells each with exactly the haploid number of chromosomes that are genetically homologous.

The bivalents in a normal first metaphase vary considerably in appearance because of the influence of various factors. As mentioned, the number and position of chiasmata is of considerable importance, and so is also the localization of the centromere, the degree of spiralization, and the size of the chromosomes themselves.

The structure of the bivalents can most easily be studied in a sideview of the first metaphase, when the chiasmata are in the equatorial plane. If all chiasmata are fully terminalized, the bivalents with two chiasmata will form rings, and those with a single chiasma form rods. Both may, however, vary slightly in form depending on the situation of the centromere. When some of the chiasmata are interstitially located, the bivalents will produce variously complicated pictures which also are affected by the size of the chromosomes and the place of their centromeres. Bivalents of sex chromosomes are always rod-like and heteromorphic in the male meiosis of dioecious higher plants, and ring-like in the female meiosis.

As already mentioned, the chromosomes are the sites of or perhaps rather made up of the genes, or of DNA chains if the reader so prefers; their effects are studied by geneticists requiring experimental facilities, whereas their chemistry and function are studied by molecular geneticists needing an electron microscope and applying the methods of physics to their investigation. It is the subtle variation between these chemical units that is the basis of the diversity of the living world. Constancy and permanence of these basic units of life is of an immense importance; to ensure their exact replication mitosis and meiosis were developed early in the history of life and these processes penetrated from the lowest to the highest eukaryotic biota. No biological process has ever been more exact or more conservative.

Although the genes are extremely stable, they are not entirely unchangeable, and their diversity is fundamental for the variation in the living world. This nature has taken care of by building into the chromosomes certain possibilities of errors or accidents, which change genes and produce new ones. This we call mutation. Mutations are constantly being formed in a very low frequency, so low that they are rarely observed even in the most common of organisms, though actually all genetic variation descends from this process. Mutations may be formed by hits of invisible radiation or by some chemicals, but most frequently they are the results of minor mistakes during the pairing and replication of the chromosomes, usually at meiosis. Most of these accidents consist of a loss of material or of an alteration in composition or arrangement of the DNA that makes the gene recessive or inactive; such changes are steps backward and detrimental to the individual and population, so a chromosome thus maimed will be at a disadvantage and sooner or later be lost from the gene-pool, which is the total gene information encoded in the entire collection of genes in a breeding population or a biological species. An infinitely low frequeny of these rare accidents, however, will be mutations that cause changes which add new material to the genes or strengthen them in some respect or another through recombination of their DNA or by so-called gene conversion, forming, for instance, what we call codominant, pleiotropic, or polymeric genes (cf. Pandey 1972).

Such constructive mutations in the conservative chromosomes are the main source of increased diversity and the basic units of variability, but their real effects become observable only after they have had a chance to combine and recombine with other genes. This is done through hybridization, which first is restricted to the local population, or deme, which is the basic unit of evolution in which all new ventures are tried out before being allowed to mix into other parts of the gene-pool. Although it is believed that natural selection will decide about the viability or vitality of such new combinations, we should not underestimate the influence of so-called genetic drift (Wright 1921), or random fluctuations in gene frequency caused by the limiting effect of small demes on the possibilites of recombination. However, selection will take care of refining and polishing the new product so that it will fit the environment perfectly or nearly so.

These essentially genetic processes produce and shape the diversity of the gene-pool. They are basic for the development of clinal variations, which are formed by selection connected with some gradation in the environment. They also cause other local differentiation even when the populations appear to be continuous, as increasingly refined methods of analysis have revealed in recent years, so that even apparently homogeneous demes are found to be somewhat heterogeneous assemblages of minor demes. By aid of other extraneous or geographical discontinuities these processes also shape parts of the gene-pool into minor geographical races, which the botanists call varieties, and into major geographical races, which are the subspecies of the taxonomists. That, however, seems to be the end of the genetical evolutionary line, or of genetical adaptations in the strict meaning of the term, or of the subspecific change, because other processes appear to be required to form a new gene-pool which will be effectively and permanently isolated from its relatives by aid of internal reproductive barriers which are characteristic of the biological species. If this conclusion based on experimental observations by geneticists is correct (cf. Müntzing 1930; Vickery 1959; Löve 1964; Gustafsson 1973), then at no level are races of the gene-pool to be regarded as being incipient species, as was believed by Darwin (1859) and is still the opinion of many evolutionists (cf. Dobzhansky 1970), for the simple reason that the creation of interfertile races requires an undisturbed or at least only a slightly disturbed miscibility, so that whenever new conditions arise they will be able to escape back into the original gene-pool with which they always share the great majority of their homozygous gene combinations, whereas species are characterized by distinct reproductive isolation.

It ought not to be overlooked in this context, that although barriers to gene exchange have not been produced between major geographical races that have been spatially or otherwise isolated for an infinite number of generations, such intraspecific taxa have sometimes been found to have formed complex homeostatic gene combinations for close adaptation to their special environment. When these compounds are fractured through hybridization, this in some cases causes various genetical disruptions which may be perturbing to the vitality of the first generations, or until a new balance has been reached by the aid of natural selection. This does not seem to directly affect the possibilities of undisturbed gene flow between such races of the same biological species, and frequently superior new combinations of a great agricultural importance have derived from just this kind of hybrids. In the great majority of hybridiza tions between widely different minor and major geographical races, however, the first generation hybrids are more vigorous than either parental race, although this tends to level off by aid of natural selection in later generations. Nevertheless, sometimes the superiority seems to prevail or aberrant forms occur, for reasons unknown. The complex question of racial hybrids is among the many cytogenetical problems awaiting a full and detailed investigation for which plants may be better suited than animals.

The chromosomes have another built in mechanism which affects their constancy above the level of the gene and shuffles around parts of chromosomes with blocks of genes without causing gene mutations. By aid of an accident of a greater magnitude, moderate to extensive parts of one or two chromosomes may break loose and then either remain broken or unite again in a way different from what formerly was. This is called fragmentation. Such a breakage occurs very rarely in nature as a result of certain disturbances in the meiotic division or because of the age of the mitotic chromosomes themselves. But it can also be induced artificially by X-rays, ultraviolet light, temperature shocks, antibiotics, various drugs, biocides, pesticides, mustard gas, artificial colors, some food additives, heavy metals and numerous other chemicals, and even by common viruses (Kihlman 1966). Naturally, most changes of this kind tend to be harmful to the well-being of the individual or even lethal, because they disturb its delicate genetical balance, and most frequently the broken chromosomes will soon be selected away. However, plants or demes which succeed to become homozygous for deviations from the usual chromosome morphology may survive, and then such changes can become important for the conservation of certain genetical combinations and for evolutionary progress. We will discuss alterations of chromosome number and their effects on a later page.

The simplest and most common breakage of chromosomes is connected with the formation of chiasmata, when two chromatids in homologous chromosomes break and unite again to exchange parts with each other. The chiasmatype theory (Janssens 1909; Darlington 1937, 1965) is basic for our understanding of meiosis and the explanation of linkage and several other genetical phenomena. These breakages, however, are normal and do not usually result in any unexpected changes, though this sometimes apparently happens through some mistakes in the pairing or recombination of the broken chromatids.

Breakages of a different extent may occur in entire chromosomes, and then include both the chromatids, or only a single chromatid. If it occurs in one place only and does not heal again at once, then it may separate a small or large part of a chromosome arm without a centromere from the chromosome itself, resulting in a genetic deficiency (Bridges 1917) and the making of an acentric fragment which will soon be eliminated, usually during mitosis. Such a fragment can also be formed from an intercalary part of the chromosome through two breaks after which the other parts of the chromosome unite and heal. That is called deletion (Painter and Muller 1929). The chromosome so shortened will pair with its partner at meiosis and form an unequal bivalent (Fig. 16b), but the fragment without a centromere will remain unpaired.

When a segment of a chromosome is doubled following a break, this is called duplication (Bridges 1919; Fig. 16c). Duplications may be small or

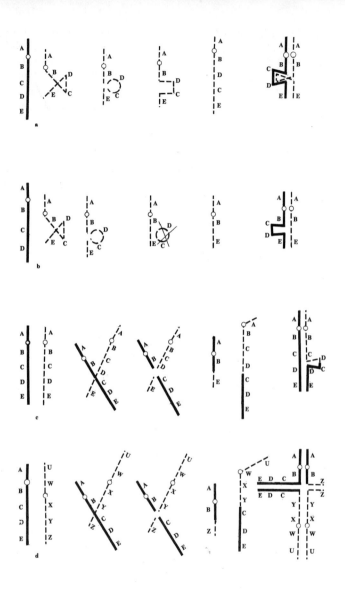

Fig. 16: Linear chromosome aberrations. a. Inversion. b. Deletion. c. Duplication. d. Segmental interchange. Based on Sharp (1943).

large; they increase the genetical material and impede the pairing at the first meiotic prophase. Usually the segment is contained within the same chromosome, although sometimes it is translocated or relocated onto a non-homologous chromosome, or, in case it includes a duplicated centromere, it may form a new but smaller chromosome which later could be duplicated again and thus change the basic number of the taxon through dysploidy.

Another simple kind of change involves chromosomes that for some reason are left unpaired at the first meiotic division. Their centromere tends to misdivide and split crosswise instead of lengthwise to break the chromosome into two halves which go to opposite poles. In this way, each arm of the old chromosome will become a new and independent chromosome with a terminal centromere; it will become a telocentric chromosome. Such chromosomes are rarely stable, however, because their centromere is weak and may soon fail to divide so the undivided chromosome passes to one nucleus and forms another new chromosome with identical arms. This is called an isochromosome (Darlington 1939). It is a kind of a large duplication, and it is new in shape and changes the genetical balance and the basic number and may lead to the evolution of a new generic line, when the isochromosome has stabilized by changing its arms genetically in one way or another.

Sometimes the chromosomes break at two points in the same nucleus at about the same time. The four broken ends then may unite again, either as they were or in a new way. If the breakages were on the same chromosome, the middle segment may become inverted before it unites with the other parts to change its genetical sequence (*abcdef* becoming *abdcef*); that is called inversion (Sturtevant 1926; Fig. 16a). Or, if the breakages were on two chromosomes, the segments may be translocated so that the chromosomes *AB* and *CD* become *BC* and *AD*; that is called segmental interchange (Belling 1927; Fig. 16d). Such chromosome changes, which include only the linear rearrangements of the genes, tend to become homozygous so the structural diversity becomes evident only after hybridization of individuals or demes differing from each other in this way. Structural hybridity causes sterility, and when these changes accumulate in different demes through hybridization and selection, they may successively build up meiotic disturbances which result in reproductive isolation.

Segmental interchange, which is characterized by a shift in position of chromosome segments within the chromosome complement, may be of different kinds that affect the pairing at meiosis in various ways. Internal segmental interchange is the change in position of a segment within the same chromosome; it is said to be extraradial if it is from one chromosome to another, but infraradial if it is within the same chromosome arm. Most segmental interchanges are, however. interchromosomal, involving the transfer of a segment from one chromosome to another or a reciprocal exchange of segments between chromosomes. If the chromosomes are homologous, the exchange is said to be fraternal, but external if the chromosomes are non-homologous. Terminal translocations of chromosome parts to the natural ends of another chromosome to not occur. Segmental interchanges may involve the chromosome itself, a single chromatid, or one of the units of a chromatid. The points at which they occur is called interchange site or point. Such points may be at any part of a chromosome.

Segmental interchanges may be either asymmetrical in which one dicentric chromosome and one acentric fragment arise, or symmetrical when both the interchange chromosomes are monocentric, with one centromere each. Asymmetrical interchanges are rare or difficult to discover; at anaphase they may form a chromatin bridge and a fragment if the centromeres of the dicentric chromosome are distributed to opposite poles, exactly as is the case of certain inversions to be discussed below.

In structural hybrids heterozygous for a symmetrical segmental interchange, no two chromosomes are homologous but four share partial homology. Consequently, at pachytene the pairing will result in the forming of a crosslike structure involving all the chromosomes (Fig. 16d), because that is the only way in which such chromosomes can pair at their entire length. The later course of this configuration depends upon the location and frequency of the chiasmata and the orientation of the centromere. If chiasmata are formed in all the four interstitial segments, then a simple ring of four chromosomes will be formed at diakinesis and metaphase I. If the formation of chiasmata fails in one of the pairing segments, a chain of four chromosomes is produced, but if the chiasmata are absent from two adjacent or alternate segments, a chain of three and a univalent, or two bivalents, will be formed. Both the rings and the chains will contain structurally normal and structurally changed chromosomes in turn. If a chiasma is, however, formed in one of the segments, simple disjunction is no longer possible so that considerable cytological and genetical irregularities occur and cause various degrees of sterility of the gametes. The occurrence and length of the interchanged segment is evidently of importance for the behavior and fertility of simple interchange hybrids, because its length clearly affects the occurrence of chiasmata.

The separation at the first anaphase of the four chromosomes in a ring or chain configuration is determined by the orientation of the centromere, which is either concordant or discordant. If the orientation is concordant and the chromosome distribution is alternative, those chromosomes which are arranged alternatively in the configuration go to the same pole, whereas adjacent distribution implies that neighboring chromosomes move to the same pole. In the former, a complete haploid chromosome set will reach each pole and form a normal pollen grain or egg cell. The same happens in the type of adjacent distribution when neighboring chromosomes in the ring or chain go to the same pole because they have homologous centromeres, whereas when neighboring chromosomes with non-homologous centromeres reach the same pole, this results in genetical duplications and deletions and in genetically unbalanced gametes. This is the reason for the so-called semisterility of plants with segmental interchanges in which about 50% (30-70%) of the pollen grains may be abortive and empty.

The above description assumes that the centromeres of the ring or chain have been oriented concordantly so that the same number of chromosomes will reach either pole. In the case that the orientation of the centromeres is discordant, however, gametes with $n + 1$ or $n - 1$ chromosomes will be produced, a result comparable to occasional non-disjunction of bivalents when for some reason both chromosomes go to the same pole. Fertilization of normal eggcells by such deviating gametes will result in the formation of either monosomic $(2n - 1)$ or trisomic $(2n + 1)$ individuals, the latter being called tertiary trisomics or translocation trisomics if they contain one or both chromosomes which took part in the segmental interchange (cf. Khush 1973).

Sometimes one chromosome of a homologous pair may take part in two or more segmental interchanges, causing the formation of configurations of six or more chromosomes at meiosis. Such structural hybrids rarely show a lower frequency of abortive pollen than 50%, and frequently they are considerably more sterile.

Fig. 17: Permanent segmental interchange heterozygote in *Oenothera biennis* ssp. *erythrosepala* (= *O. lamarckiana*). a. $1_{XII} + 1_{II}$. b. Explanation. After Cleland (1972).

In some plants, selection has refined a kind of a permanent structural hybrid in which the segregation of the multivalents behaves as if each of the two chromosome complexes were a single unit. These are the so-called complex heterozygotes, in which several or all the chromosomes are linked in a ring or chain at meiosis as a result of segmental interchanges. The chromosomes concerned associate at meiosis end-to-end in distinct rings, the size of the ring and the number of chromosomes in it being constant for each taxon, except that rarely a chain or two chains could replace a larger ring. The ring of chromosomes usually is arranged so that adjoining chromosomes go to opposite poles (Fig. 17). It has been shown that the chromosomes have a fixed position in the configuration so that the chromosomes of one parental gamete pair at opposite ends with ends of two chromosomes from the other gamete, and at anaphase the two original parental combinations are the only ones reproduced or viable. The group of chromosomes and genes of each unit is called a complex; it acts as a single linkage group. Permanent structural hybridity has been studied most intensely in *Oenothera* (Cleland 1972), but it occurs also in some other plants, as, e.g., *Rhoeo spathacea* (Lin and Paddock 1973).

Inversion is a structural change of the chromosomes involving a reversal of a segment and its linear gene arrangement relative to some other arrangement of the chromosomes or linkage group in question. It always involves an

48

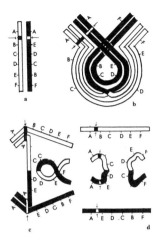

Fig. 18: Inversion loop at prophase I and its results at anaphase I. a. Heterozygosity for a paracentric inversion. b. Pairing in a loop at pachytene. c. Inversion bridge and fragment at anaphase I. d. Telophase. After Dobzhansky (1970).

interstitial part of the chromosome and requires two breaks and a complete reversal of the segment, since there is no evidence for the occurrence of terminal inversions.

Several kinds of inversions have been recognized, but they are either single or complex. If a single inversion is between two breaks in the same chromosome arm, it is said to be asymmetrical or paracentric, but if the breaking points are located in different chromosome arms so the inverted segment includes the centromere, the inversion is said to be symmetrical or pericentric. Complex inversions, however, are of four different kinds. If inverted segments are separated by an unchanged part of the chromosome, the inversions are said to be independent and may well be the results of breaks that occurred independently at different times (*abcdefgh* → *acbdegfh*). If two inverted segments border to each other (*abcdefgh* → *acbedfgh*), we speak about direct tandem inversions, whereas reversed tandem inversions include two inverted segments which are adjacent but

49

Fig. 19: Bridges and fragments caused by paracentric inversions in *Fritillaria* at anaphase-telophase I. After Bennett in Sturtevant and Beadle (1940).

mutually interchanged (*abcdefgh → aedbcfgh*). Included inversions have an inverted segment within which a smaller part is inverted again (*abcdefgh → agfedcbh → agfdecbh*). At last, inversions are said to overlapping, when a part of a segment is inverted a second time together with a part which was not inverted the first time (*abcdefgh → aedcbfgh → aedgfbch*).

Any type of inversion, whether including the centromere or not, will cause a complication of the meiotic division of an individual heterozygous for it, because it results in the reverse pairing of a loop at pachytene (Figs. 16a, 18). The pairing behavior of the inverted chromosome and its structurally unchanged partner depends on the length of the inversion and the longitudinal relationship of the inverted and unchanged chromosome segments. It is in a long inversion that the pairing requires the formation of a backward loop in the unchanged chromosome so that the inverted part of the other chromosome can pair its homologous loci exactly with it by forming a forward loop. If the inverted segment, however, is so short, that loop formation is unattainable, the segment may either remain unpaired, or nonhomologous segments may seemingly unite. In the case of a very long inversion, it may even pair by reversing the entire chromosome without forming a loop, leaving the shorter terminal segments unpaired.

Further influence of inversion hybridity on the meiotic division is connected with the formation of chiasmata inside and outside the inverted segment. If the heterozygous inversion is paracentric, a chiasma within the loop will cause the formation of a dicentric (with two centromeres) chromatid and an acentric fragment. The dicentric chromatid makes a bridge at anaphase in the first meiotic division and a loop chromatid in the second anaphase, but the acentric fragment is passive and may soon be eliminated (Fig. 19). Two chiasmata formed within the loop may result in a double bridge and two fragments, or they may produce a monocentric loop chromatid with both ends joined at the centromere and a single fragment, the loop then very likely forming a bridge at second anaphase. And, finally, three chiasmata within the loop are likely to produce two monocentric loop chromatids and two fragments which result in two bridges with fragments at second anaphase. In the case of heterozygous pericentric inversions, chiasmata do not cause the formation of dicentric chromatids, loop chromatids, or fragments, but they may cause duplication and deletion of the terminal segments.

Inversions, segmental interchanges, and duplications may survive in the population without appreciable effects, although in hybrids they will cause sterility which disturbs genetic segregation, but even that may be tolerated for a long time, especially in smaller demes in which they are more common than in larger populations(Gustafsson 1972, 1973). However, sooner or later such hybridization increases the chromosomal rearrangements through additional breakage because of mechanical disturbances at meiosis, especially the inversions. By time, further hybridization accumulates these changes in the demes that become homozygous for them and then become increasingly independent from other demes because of limited interfertility and increased incompatibility and other factors leading to somewhat separate variation. Such populations, which may or may not be morphologically distinct at this stage, are the real incipient species, or perhaps rather semispecies as long as their isolation is incomplete (Legendre 1972), and they are the most frequent causes of confusion among taxonomists for various reasons. Slowly but inevitably such an accumulation of chromosomal rearrangements will lead to what is called gradual speciation (Valentine 1940; Löve 1964), or the gradual creation of an effective reproductive isolation between the new and the original gene-pools. It is reproductive isolation of such an internal kind involving the chromosomes that is the most fundamental characteristics of the biological units which are traditionally called species, or of the evolutionary stage when gene exchange ceases and two populations become effectively launched into different channels of evolution.

Gradual species sometimes commence allopatrically (fr. Greek *allus*, another, and *patra*, native land, Poulton 1903; Mayr 1942) as the result of an accidental but effective geographical isolation, rarely by a long-distance dispersal of a small founder (Mayr 1942) deme but more often by the drifting apart of the land with a fraction of the population; the process of diversification is, then, extremely slow. More frequently, however, this kind of speciation starts sympatrically (fr. Greek *sym-*, with, together) within a deme and is then most effectively accelerated when parapatric demes differing in chromosomal rearrangements meet and hybridize and accumulate new breakages on a grand scale (Lewis 1966). If the new arrangements are deleterious, they will soon disappear. Sometimes they may succeed to avoid complete isolation by developing into permanent heterozygotes, though even these are ultimately a dead-end because they limit diversity. But if and when these new taxa get past all the dangerous hurdles on their way towards homozygosity of the advantageous new chromosome combinations, then they have formed a new gene-pool and a new species and sent it off on its way to distinct diversity of its own. With other words, species are not primarily the result of morphological diversity but of reproductive distinctness, and first after such isolation has been realized are the then distinct gene-pools able to differentiate morphologically and independently.

The gradual speciation process, or the course of accumulation of chromosomal rearrangements, is the common means of evolution of reproductive barriers that isolate gene-pools. There is, however, still another chromosomal mechanism which leads to the most effective reproductive barrier known and produces it instantly. This is the so-called abrupt speciation (Valentine 1949; Löve 1964), which by aid of polyploidy doubles the chromosome number of a hybrid or of an individual or a deme in a process equally irreversible as is gradual speciation. Polyploidy is relatively common in the higher plants (Löve and Löve 1949, 1971), the algae (Godward 1966), the fungi (Rogers 1973), and certain animals (White 1973). The process has been described on an earlier page, but we will look closer at the effects of polyploidy on meiosis and on sterility in the next chapter.

C. Meiosis in polyploids and aneuploids

The meiotic processes described in earlier chapters are those forming the gametophyte of a diploid spermatophyte. The course of meiosis is essentially the same in panallotetraploids, which behave like diploids and form only bivalents at first prophase and first metaphase because every one of their chromosomes is only represented twice.

In aneuploids, monosomics will constitute no problem, since their single chromosome of the pair in question will have no partner to pair with and therefore can only form a univalent. Such univalents frequently go through the first division undivided, although sometimes their centromere may split prematurely at first instead of second anaphase, and sometimes the centromere splits transversely to form two telocentric chromosomes which may become isochromosomes already in the following division. Trisomics, however, will pair with two partners during prophase when possible. Since the chromosomes always come together in pairs, but different pairs are formed at different points, the three homologous chromosomes may easily form a configuration of three, which we call a trivalent. Trivalents behave much like a bivalent during the prophase, but at diakinesis and metaphase of the first division they may form a ring or chain of three or a panlike configuration, depending upon the localization of the chiasmata and their terminalization which is essentially similar to that of ordinary bivalents of the same species. At first anaphase, two of the chromosomes will most frequently go to one pole, one to the other, although sometimes the struggle between the centromeres may cause that all three go to the same pole. But sometimes one of the chromosomes may remain unpaired throughout the meiotic division.

In autoploids and hemialloploids, and also in autoalloploids, the situation is more complex because of the higher number of identical chromosomes. Panautoploids have several complements of largely or fully identical chromosomes, all of which could pair and form as many multivalents as there are chromosomes in the monoploid complement or basic number of the taxon. However, such a complete pairing rarely occurs, and usually only about half of the possible number of multivalents is formed. There is experimental evidence from polyploid *Triticum, Avena* and *Nicotiana* in support of the hypothesis that bivalization or reduction of multivalents in autoploids and hemialloploids may be conditioned by special genes (Rajháthy and Thomas 1972). Also, in mosses, there seem to be intrinsic

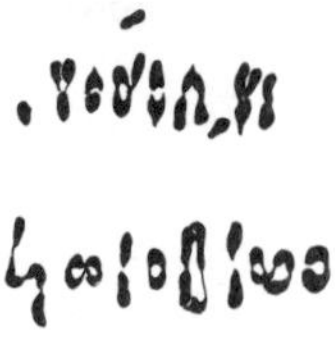

Fig. 20: Multivalents at meiosis in *Dactylis glomerata*. After Müntzing (1971).

factors that operate against multivalent formation (Anderson and Lemmon 1972). Smallness of chromosomes always reduces multivalent formation because it reduces the frequency of chiasmata (Darlington 1937), so that the absence of multivalents is not necessarily an indication of alloploid origin, as has sometimes been suggested.

Multivalents in panautotetraploids are most frequently quadrivalents that take upon themselves various forms as closed or open configurations due to diversity of chiasma formation, but trivalents and univalents are also frequent (Fig. 20,21). Furthermore, the irregular orientation of the centromeres of these multivalents at the first metaphase of meiosis greatly disturbs the normal distribution of the chromosomes at first anaphase, resulting in nuclei with deviating chromosome numbers which may be either inviable and thus sterile, or give rise to aneuploid offspring with reduced vitality. In this respect, hemiautoploids and hemialloploids are intermediate and form an increasingly lower frequency of multivalents and, thus, a lower degree of sterility approaching that of panalloploids and diploids.

In diploids and polyploids with multivalents, it is customary to report the total number of configurations of chromosomes at diakinesis or metaphase I in a simple formula. For instance, if there are 14 diploid chromosomes that form seven bivalents at metaphase, the configuration is simply 7_{II}. If there is a segmental interchange, the configuration in one cell may be 7_{II}, or seven bivalents, in another $5_{II} + 1_{III} + 1_I$, or five bivalents, one trivalent, and one univalent, and in several cells it may be $5_{II} + 1_{IV}$, or five bivalents and one quadrivalent. In a species of *Oenothera* there may be $1_{II} + 1_{XII}$, or just 1_{XIV}. An autoploid with 2n = 28 chromosomes could have 7_{IV} or 14_{II}, although intermediate configurations, as, e.g. $8_{II} + 3_{IV}$, would be more likely the rule. Despite that such configurations are observed during

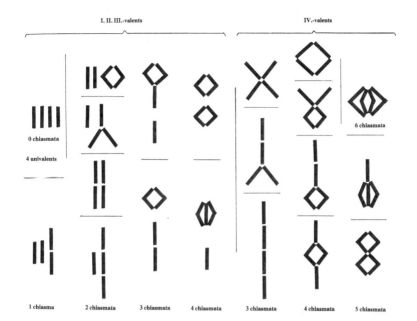

Fig. 21: Multivalents in an autoploid. After Darlington (1937).

meiosis, it should be understood that they have nothing to do with either n or $2n$ chromosome numbers, although counted together all the chromosomes of a total configuration will match the diploid complement of the plant. To report such configurations as, e.g., $2n = 8_{II} + 3_{IV}$, in order to tell at the same time about the chromosome number and the occurrence of these meiotic disturbances, is actually only confusing and hardly correct and so ought to be avoided in reports of scientific observations.

Triploids, pentaploids, or other uneven euploids which are formed through hybridization between taxa with euploid numbers, may behave at meiosis like trisomics and autoploids, if their parents have essentially homologous chromosomes. If the tetraploid parent of a triploid, however, was a panalloploid with only one chromosome complement identical to that of the diploid parent, the hybrid will form one set of bivalents and one set of

Fig. 22: The origin of tetraploid *Triticum turgidum* (AABB) and hexaploid *Triticum aestivum* (AABBDD), by alloploidy from the haplomes of diploid *Triticum monococcum* (AA), *Triticum speltoides* (BB), and *Triticum tauschii* (DD).

univalents. Such triploids (or pentaploids when the parents were tetraploid and hexaploid, etc.) have been used in so-called haplome-analysis of polyploids in order to reveal their original cytological relationships (Kihara 1930), while genome-analysis reveals their genetical relationships.

The basic assumption of the haplome-analysis is that the essentially identical haplomes or monoploid sets of chromosomes of alloploids and their diploid relatives will remain identical or so slightly changed for a long period of time that their individual chromosomes will pair normally in artificial hybrids. A triploid hybrid between an allotetraploid, $AABB$, and either of its diploid progenitors, AA or BB, is expected to form $2x$ bivalents and x univalents, or $X_{11} + X_1$, the latter then representing the chromosomes of the progenitor that does not take part in the hybridization. The method has been especially successful in identifying various haplomes of the wheatgrasses and then especially those of the cultivated wheats themselves. The cultivated wheats include three ploidy levels, which represent three well-defined biological species each of which including fully interfertile subspecies or cultivars, which are supposed to have been formed from the original species by selection during the thousands of years of their cultivation. After hybridizing the diploid *Triticum monococcum* L. to the tetraploid *Triticum turgidum* L. as well as to the hexaploid *Triticum aestivum* L., the triploid hybrid was found to have the meiotic configuration $7_{11} + 7_1$, but the tetraploid hybrid had $7_{11} + 14_1$. From this could be concluded that the haplome of the diploid species, designated A, must be met with in both the other species. Likewise, the pentaploid hybrid between the tetraploid and hexaploid species had the configuration $14_{11} + 7_1$; since one of the paired haplomes must be A, another haplome, named B, must be met with in both the tetraploid and hexaploid taxon, whereas the univalents represent a haplome, called D, that is only met with in the hexaploid (Fig. 22).

A search for the two additional haplomes by aid of morphological and experimental observations revealed (McFadden and Sears 1946) that the donor of the D haplome is the related taxon *Triticum tauschii* (Coss.) Schmalh. ($= Aegilops squarrosa$ auct.). This could be verified by aid of hybridization of this weed with the tetraploid species of wheat in which hybrid no pairing was met with because this haplome is absent, and with the hexaploid wheat which resulted in plants with $7_{11} + 14_1$ as expected. The final verification came, however, through the production of an artificial hexaploid from the triploid hybrid between the tetraploid *Triticum turgidum* and the diploid *Triticum tauschii*. That hexaploid is morphologically very close to *Triticum aestivum* and when these hexaploids are hybridized, they form mainly bivalents at meiosis.

Several dead ends were tried in the hunt for the B haplome which evaded discovery until Sarkar (cf. Sarkar and Stebbins 1956) applied the so-called pictorial scatter diagram method of Anderson (1949) in the search, after first having excluded numerous earlier proposals by cytological observations of the karyotypes. This method is based on the understanding, that, by plotting on a diagram the characteristics of a hybrid and its known parent, the cluster of characteristics found to be additional in the hybrid will constitute a kind of a description of the unknown parent, a reasoning that also could be made for alloploids which are of hybrid origin. From such studies it was concluded that the most likely donor of the *B* haplome would be the species *Triticum speltoides* (Tausch) Gren. or its close relatives, a suggestion later supported by aid of haplome analysis by several other workers. It is known from archaeological research that the union of the first two haplomes took place near the eastern end of the Mediterranean about 10.000 years ago, whereas the third haplome was added almost 8000 years ago where the tetraploid species was cultivated. The haplomes have been together in the same cells for as many generations, because all the species are annuals. All three haplomes still remain essentially unchanged, although some minor chromosomal rearrangements involving exchanges between haplomes have taken place, especially in the tetraploid species, since hybrids between the original and artificial polyploids sometimes form some configurations at meiosis that indicate heterozygosity for additional inversions and segmental interchanges. That is one more attest to the claim of the extreme conservatism of the chromosomes.

The haplome-analysis has been a valuable tool in the search for new material and new methods for improving some cultivated alloploids. The survival of man depends largely on his ability to understand and govern the cytogenetical mechanisms basic to the evolution and improvement of plants. Although the classical methods of breeding have utilized hybridization and sophisticated selection with considerable success to improve the already cultivated material, more advanced practices of recent decades apply artificial auto- and alloploidy and all kinds of cytological manipulations and artificial chromosomal rearrangements by aid of X-rays and chemicals in the attemps to produce new cultivated plants and increase and improve the diversity of the older ones. A concise introduction to these methods is given by Lawrence (1968).

Theoretically, it is possible that the meiosis of a triploid could be so badly disturbed that all the chromosomes became included in a single nucleus, a monad, or that all the chromosomes of one parent could by haphazard go to one dyad and all those of the other parent to another, resulting in normal and fertile gametes with the haploid, diploid and triploid chromosome number. In reality, however, the meiosis of a triploid or an unevenly euploid hybrid is usually so profoundly disturbed that all the gametes receive intermediate and aneuploid numbers and become so unbalanced that they do not develop. If, however, some should be viable, they will give rise to a highly disturbed next generation which is doomed to disappear.

Observations of this kind are the actual basis of the claim, that the occurrence of two evenly euploid chromosome numbers in a series of related taxa constitute the demonstration of a presence of the most effective reproductive barrier known, because after hybridization between two ploidy levels the offspring will inevitably have an uneven multiple of the basic number, and neither parent combination will ever be recovered in later generations, if any new generation at all is being produced. Therefore, all gene-flow will be effectively prevented either through the complete sterility of the F_1 hybrids themselves, or, if a low degree of fertility should be maintained and a following generation made possible either through selfing or through backcrossing to one of the parents, these individuals will show low viability and high sterility. Crossability certainly indicates some relationship, but something else is responsible for the fact that often there is a complete incompatibility or crossing barrier between a diploid and its panautotetraploid, whereas this barrier breaks down between the tetraploid and hexaploid and, still more, between the hexaploid and octoploid in the same panautoploid series. Since such hybrids are, nevertheless, sterile and unable to mix freely and form viable later generations or reproduce the parental combinations, the ability to form an F_1 hybrid means nothing in a discussion of reproductive isolation, because it is not crossability but miscibility which is characteristic of various races of a species. And really fertile hybrids between two levels of even artificially produced panautoploids have never been observed.

This carries us to the statement, only indirectly made earlier, that every well defined biological species, which is reproductively isolated from other such taxa (Mayr 1942), is characterized by a single chromosome number only. This was originally conceived by Wilson (1900) and has later been verified by thousands of critical investigators working with plants, animals, and fungi. This notwithstanding, there are species which taxonomists of the past have given such wide delimitation, that cytologists find them to include taxa with more than one chromosome number. Naturally, this does not change the validity of the above claim, but only indicates to the observer that the taxonomical species in question has not been circumscribed according to the biological definition of the category, and that a close and critical search for morphological differences must be made in order to facilitate the identification of the biologically isolated species that are characterized by the chromosome numbers observed. In some cases, especially those involving autoploids, such an investigation might show that only a few differentiating characters are met with. But although it is a common assertion by some observers, that certain plants with different chromosome numbers are completely identical, this may seem to be a genetical absurdity, especially in view of the fact that even strictly panautoploid demes of the purest lines of cultivated autogamous barley are morphologically distinguishable. We have made a closer check of hundreds of such claims of complete identity of two or more ploidy levels without ever being able to verify the correctness of this, when the comparison has been detailed and exact, although in some cases the differences are admittedly few and small. Instead we have frequently found that some skilled taxonomist long gone had actually observed morphological distinctions between such taxa without knowing about their chromosome numbers, and described them taxonomically at or below the level of species, although his conclusions may have been ignored by later colleagues compiling flora manuals.

These are questions that have been vigorously debated in the past, and were, and still are, loaded with emotion. They will not be solved through passionate arguments but by the practicability of the results achieved. But during the stage of controversy perhaps either side may find some consolation in the expression by the American botanist Fernald (1946) in a case where he actually but unwittingly was wrong and his antagonist right: "Inability to see what others clearly see is not a sin; neither does it necessarily clarify a question."

Fig. 23: Chromosome number increase through agmatoploidy in *Luzula*. a. *L. spadicea*, 2n = 12. b. *L. parviflora*, 2n = 24. c. *L. multiflora*, 2n = 36. d. *L. sudetica*, 2n = 48. After Nordenskiöld (1951).

D. Meiosis in agmatoploids

Normal meiosis reduces the number of centromeres and chromosomes during the first division, whereas the second division is equational and essentially mitotic. The situation is different in agmatoploids, which are plants or animals with polycentric or holocentric chromosomes in which each chromatid seems to have either numerous centromeres or a centromere active in the spindle throughout their entire length so they act like completely autonomous and independent structures. Because of this centromere arrangement, the chromosomes of an agmatoploid may divide crosswise to create an agmatoploid series (Fig. 23) and form new chromosomes which at meiosis may pair with parts of the longer chromosomes. In such plants, the relation between the two divisions is reversed from the condition of a normal meiosis, so the first meiotic division is essentially mitotic and results in the unreduced number of chromosomes after the first anaphase. During interphase a reassociation of two homologous chromatids occurs so the chromatids are paired at second metaphase, and the actual reduction of the chromosome number takes place at the second anaphase (Fig. 24). Agmatoploidy is met with in certain members of the Cyperaceae and Juncaceae and also in some insects and algae. It may be complete so that all the chromosomes seem to have broken transversely at the same time, as is common in *Luzula* (Nordenskiöld 1951), or partial so that only some of the chromosomes are so divided simultaneously. In some genera, as, e.g. *Carex*, partial or complete agmatoploidy is frequently combined with true polyploidy,

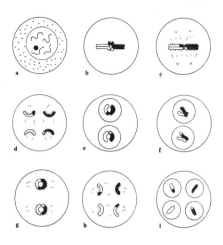

Fig. 24: Meiosis in an agmatoploid with polycentric or holocentric chromosomes. a. Leptotene. b. Diakinesis. c. Metaphase I. d. Anaphase I. e. Telophase I. f. Prophase II. g. Metaphase II. h. Anaphase II. i. Telophase II.

causing widely dysploid basic numbers even within the same taxon (Löve, Löve and Raymond 1957; Faulkner 1972). The mechanism of agmato-ploidy and, especially, its meiosis are only superficially known.

E. Diaspore formation

a. Fertilization and seed formation

In order to survive and fill the earth, plants form organs by aid of which they move from one place to another or, in the cases of annuals, from one season to another. The process of plant movement is called dispersal (Ridley 1930; van der Pijl 1969). It is a one-way change of residence per lifetime as contrasted to the seasonal migration of animals, and the organs with which the move is accomplished are collectively called

Fig. 25: Diagram of a normal fertilization of an embryo-sac. After Müntzing (1971).

diaspores (fr. Greek *diaspeiro*, I broadcast, Sernander 1927). Although other organs of higher plants are involved in dispersal, by far the most frequent diaspore is the seed. which contains a temporarily dormant embryo and generally a special nutritive tissue, surrounded by a seed coat formed by the integuments of the ovule from which the embryo originated. The growth of the embryo generally starts with fertilization of the egg cell and the polar nuclei by the sperm nuclei; the egg cell and the sperm nuclei have attained their haploid status by aid of meiosis.

The purpose of the meiotic division is the production of the gametophyte generation in which the chromosome number is reduced to the haploid number (*n*) and the genes in the chromosomes reshuffled in various ways to increase diversity. The male and female gametes which are formed by the gametophytes need to be united to form the new sporophyte with its diploid (*2n*) chromosome number. In the angiosperms, this is done through transportation of the pollen grains, either from the anthers to the stigma of the same flower or between flowers of the same individual or between different individuals, a process which is called pollination. This is frequently achieved by aid of growth movements of the organs of the still closed flowers, or by aid of various pollinators or pollen vectors, the most evident of which being the wind, insects or even water. After pollination, the pollen germinates to form a pollen tube which penetrates down the style and into the ovule and embryo-sac, where the two sperm nuclei are released. One of these, the male gamete, fuses with the egg cell, or female gamete, to form the zygote from which the diploid embryo develops by successive mitotic divisions. That is called fertilization. But since so-called double fertilization (Fig. 25) is required in higher plants to make possible the development of the seed, the other sperm nucleus combines with the polar nuclei to form a truly triploid (= triple id)

63

nucleus which divides mitotically to produce the endosperm from which the embryo will be nourished. The endosperm is frequently without walls between the nuclei and often divides endomitotically. If the pollen originated from anthers of the same flower or the same plant as the stigma on which it germinates, this results in self-pollination or autogamy, whereas the transfer of pollen between individuals causes cross-pollination or allogamy. Both phenomena are frequent and result in somewhat different genetical systems.

Much more pollen is released than is necessary for securing fertilization, and many more grains germinate on the stigma than is needed for the fertilization of the available embryo-sacs. Therefore, a stiff competition arises between genetically different pollen tubes all of which do not grow at the same rate. That is called certation (fr. Latin *certatio*, to compete, Heribert-Nilsson 1915).

In gymnosperms and pteridophytes the process of meiosis is in general very similar to what has been described for the angiosperms, and the fertilization process in gymnosperms is essentially the same as described above. In the true ferns, however, there are certain differences in the fertilization process which are caused by the fact that in these plants the diploid sporophyte and the haploid gametophyte constitute different individuals that grow separately. Meiosis produces the usually 64 haploid spores in the sori that are formed on the leaves of the sporophyte, which is the longlived plant as we know it. The spores are all identical in true ferns and germinate into prothalli which typically consist of a green thallus attached to the soil by unicellular, filamentous rhizoids borne on its underside. The prothallus generally lives only for a few weeks, reaching at most a few centimeters in diameter and superficially resembling a simple thallose liverwort. From its upperside arise numerous so-called antheridia, which produce male cells, and archegonia, in which an egg cell is formed, all by mitosis from the haploid tissue. The egg cell is fertilized by a sperm cell, always in water and thus only when the prothallus has been wetted.

The zygote develops into the diploid, asexual generation or the sporophyte, which is structurally highly differentiated and longlived. The pteridophytes are the only spore plants that develop a root apex, a stem apex and a leaf apex, which give rise respectively to the first root, the stem, and the first leaf (cotyledon) of the sporeling, which otherwise is characteristic of the seed plants.

b. Asexual reproduction

I. General

It has long been known that plants and lower animals can reproduce vegetatively by various kinds of splitting away tissues that can regenerate an entire individual. In animals this occurs regularly only in some of the lowest groups, such as sponges, coelentherates, turbellarians and bryozoans, whereas in plants vegetative propagation is a general phenomenon in all groups, involving, in the higher plants, reproduction by aid of bulbs, bulbils, tubers, rhizomes, runners, and layers under natural conditions, and also allowing the artificial multiplication of the individual by aid of all kinds of cuttings from roots, stems or branches and even from small pieces of other tissues. By these means a single animal or plant can produce a considerable array of individuals that are of importance for its success and survival. A population so formed is identical to the parent individual in every genetical respect; it is known as clone (fr. Greek *klon*, a twig, Webber 1903), of which potatoes, rhubarb, strawberries, fruit trees and various other cultivated plants are the best examples.

Autonomous development of offspring from egg cells not fertilized by male gametes occurs in both plants and animals. In plants it is named agamospermy, agamospory or apomictosis which together with vegetative reproduction without formation of seeds is called apomixis; it will be closer dealt with below. In animals it is termed parthenogenesis (fr. Greek *parthenos*, a virgin, and *genesis*, birth, creation, Owen 1849), of which there are two forms: thelytoky (fr. Greek *thely*, female, and *tocus*, offspring, v. Siebold 1871) and arrhenotoky (fr. Greek *arrhenicus*, male, Leuckart 1857) of which the latter is a kind of sex determination through the production of haploid males from unfertilized eggs. True parthenogenesis, or thelytoky, occurs either as a facultative or seasonal condition in otherwise sexually reproducing animals, or it is obligate so that male individuals are absent from the species. Obligate parthenogenesis is a rare phenomenon which frequently is connected with survival of apparent hybrids, frequently triploid. The phenomenon has been most throroughly summarized by White (1973).

Asexual reproduction with seeds in plants seems to have been first observed in *Alchornea ilicifolia* Muell. Arg., a tropical Australian

Euphorbiaceae, by Smith (1841), who did not study the phenomenon closer. Strasburger (1878), however, showed that the asexually produced seeds of this dioecious plant arise from adventitious buds in the tissue of the nucellus, but not from the egg cell itself. Some more examples of the phenomenon were added by others during the latter part of the last century, but the study of the condition in modern sense dates back only to the works of Juel (1898, 1900) and Murbeck (1897, 1901).

Early students of asexual seed or spore formation in plants misapplied the name parthenogenesis to this phenomenon, or termed it variably as apospory, apogamy or apomixis (fr. Greek *apo*, away, absent, and spory, gamy or mixis, de Bary 1877, 1878; Bower 1886). The last word is presently used, following Winkler (1908), as a general and inclusive term for all kinds of asexual propagation in plants, as contrasted to sexual reproduction, or amphimixis (fr. Greek *amphi*, on both sides, and *mixis*, mingling, Weismann 1891). The subject has been most thoroughly reviewed by Gustafsson (1946-47), Stebbins (1950) and Rutishauser (1967), and their terminology for the higher plants is mainly followed below.

II. Asexual seed formation in angiosperms

In modern terminology, asexual seed formation in angiosperms is called agamospermy (fr. Greek *a-*, a negative prefix, *gamein*, to marry, and *sperma*, seed, Täckholm 1922). Several hundred cases of this phenomenon have been thoroughly studied in recent decades from numerous families, though most of the species in which it is known to predominate seem to belong to the Poaceae, Rosaceae and Asteraceae. Although agamospermy is most frequent in polyploids or hybrids in floras of cold regions, the largest number of agamospermous species is met with in the warm regions of the world (McWilliams 1964), where it occurs equally frequently in diploids and polyploids. Agamospermy may be obligate and absolute so that the genetical recombination system is effectively closed, but most frequently it seems to be facultative and then allows hybridization and sexual reproduction to occur on rare occasions or even predominantly. Obligate agamosperms may be characterized by even or uneven euploid or aneuploid chromosome numbers, which then are the same in all individuals of the taxon, whereas facultative agamosperms sometimes may show a wide variation in aneuploid chromosome numbers of small or large demes, as is shown especially in the grass genus *Poa,* some taxa of which may in a small area be represented by agamospermous demes with chromosome numbers of the complete aneuploid series from 2n = 82 to 2n = 147 (Löve, 1952).

When agamospermy occurs, meiosis is suppressed in the female tissues, but in the male it may take place, though it is then frequently much disturbed. The egg cell, in whatever way it develops, is inevitably diploid and should be designated as being *2n.* No fertilization is required, and the embryo always has the same genotype and the same chromosome number as the mother plant, although so-called autosegregation, or occasional recombination of or changes in the chromosomes during the formation of the agamospermous gametophyte, may result in some important variability (Gustafsson 1943). The addition or loss of single chromosomes also affects this variability (Sörensen and Gudjónsson 1946) or even restores some sexuality (Sörensen 1958). Most agamosperms develop their seed autonomously, although some require a limited stimulation from a pollen tube or even from a sperm nucleus that fuses with the polar nucleus to induce the development of the endosperm, before the non-fertilized egg cell or its substitute is able to begin the development of the embryo itself. That is called pseudogamy (fr. Greek *pseudos,* false, and gamy, Focke 1881); it may even be aroused by aid of pollen from another species.

The simplest mode of agamospermous seed formation is adventitious embryony (Strasburger 1878), whereby the embryos develop directly from the diploid tissues of the sporophyte, usually either from the ovule or from the integument, and the gametophyte stage is completely bypassed. As pointed out by Stebbins (1950), this kind of agamospermy seems to be rather frequent in tropical or warm temperate zone taxa, where it is known from several families.

Other forms of agamospermy are based on an apparently normal production of an embryo-sac, although the meiotic division is bypassed or deranged so that the nuclei remain diploid and cytogenetically identical with those of the embryo-sac mother cell. If the embryo-sac is formed directly through mitotic divisions from a nucellus or integument cell, this is called apospory (de Bary 1878; Bower 1886). It may, however, be formed from an archesporial or embryo-sac mother cell, although meiosis is omitted or so disturbed that it does not lead to a reduction in chromosome number but ends in the formation of a restitution nucleus that includes all the chromosomes of the normal sporophytic tissue; that is called diplospory (fr. Greek *diploos*, and spory, Edman 1931).

When embryo-sac cells other than the egg cell of the diploid gametophyte develop into a likewise diploid sporophyte, for instance a synergid, this is called apogamety (Renner 1916). Whereas autonomous development of an egg cell, either normal and haploid or abnormal and diploid, is often collectively termed as parthenogenesis, although only the latter is identical with strict parthenogenesis or thelytoky in animals. It would seem to be more appropriate to name the latter in plants as matrogony (fr. Greek *mater*, mother, and *gone,* offspring), whereas so-called haploid parthenogenesis is better termed gynogenesis (fr. Greek *gyne,* female, and *genes,* born, Wilson 1928), or haplospory (fr. Greek *haploos*, single, and spory, Battaglia 1947).

There are other variations of agamospermy which are less frequent and less well known. Also, there are reports of the occurrence, in experimental material, of so-called androgenesis (fr. Greek *andro*, man, and genesis, Verworn 1891), in which a haploid embryo develops from a male nucleus after inactivation or elimination of the maternal nucleus during or after fertilization. Such individuals will be haploid and contain only the haploid set of the male parent. Haploids have also been experimentally produced from egg cells after pollination with pollen of another species, or with X-rayed pollen, or by much delayed pollination (Kihara 1940). Recently, experiments have shown that when anthers are left under sterile and humid conditions on a culture medium in a temperature between 20° and 30°C for a long period, even pollen grains may develop directly into haploid plants (Melchers and Labib 1970; Nitsch and Nitsch 1970; Pandey 1973). Such haploids are of considerable interest to plant breeding because by aid of artificial duplication of their chromosome number they make it possible to produce in two steps a completely homozygous strain of both allogamous and autogamous plants, a goal that seems to be unreachable with classical breeding methods. Although such haploids are known only from experiments, it is possible that some individuals that are rarely found in natural populations may have derived from germinating pollen grains, although they may also have originated as the haploid member of a twin seed (Müntzing 1938), or they could be explained on basis of some other and less farfetched genetical accident.

III. Asexual spore formation in ferns

Apomixis other than vegetative reproduction has been reported from some algae and fungi, in which a gamete may directly give rise to a gametophyte. That was termed apomictosis by Winkler (1942), and it seems to be a very rare phenomenon, if the few reports are indicative of its frequency. In the true ferns, however, sexual reproduction without fertilization is known to be considerably more frequent than in the flowering plants (Manton 1950; Evans 1969), for reasons unknown. The mode of asexual reproduction in the ferns resembles the agamospermy of the angiosperms but it is not identical with it. Therefore, we propose for it the name agamospory. Whenever it occurs, it seems to be obligate and function similarly to diplospory in the higher plants.

In both sexual and agamosporous ferns, the early development of the archesporium is identical up to the third mitotic division in the sporangium during which eight cells are formed (Manton 1950). In sexual taxa, the fourth division usually produces sixteen spore mother cells which through normal meiosis become 64 haploid spores. In the agamosporous ferns, however, the division which should change the eight-cell archesporium into the sixteen-celled is disturbed so that although the metaphase starts and the chromosomes take their place in the metaphase plate, no anaphase is formed, but the split chromosomes revert to an interphase. When the nucleus has further enlarged and comes out of this stage, the eight spore mother cells go into a seemingly regular meiosis of which the chromosome pairs are as many as are the somatic chromosomes of the plant and the individual chromosomes are twice as many, because of a kind of an endomitotic doubling during the pre-meiotic interphase. Therefore, the apparent meiosis only restores the sporophytic diploid number and forms 32 spores all of which actually are diploid.

The spores with the unreduced somatic chromosome number look normal and are fully functional, but the prothalli developing from them are likewise diploid and devoid of archegonia but rich in antheridia. A new sporophyte is formed directly from the central tissue of the prothallus without any sexual fusion and therefore also without a change in chromosome number. There is some difference in the form of the leaves of the young sporophyte since the juvenile type common in sexually reproduced ferns is absent, but otherwise the most easily detectable difference is that the sporangia of the agamosporous ferns contain 32 spores instead of the usual 64.

Agamosporous ferns may be characterized by evenly euploid chromosome numbers, although more frequently they seem to have uneven multiples which may indicate that they are actually sterile species hybrids which are seemingly able to reproduce by spores and go through the alternation of generations, although in fact both generations are cytologically identical and free of sexuality. The chromosome number of the sporophyte of a normally sexual fern is correctly reported as $2n$, that of the gametophyte and the spores from which it is produced after meiotic reduction of the diploid number is n. But it is a highly confusing and illogical absurdity to report the chromosome number of the agamosporous ferns as "$n = 2n$" or as "n" as was done by Manton (1950) and continued by most later fern cytologists, because even their gametophytes and the spores produced by their kind of pseudo-meiosis are characterized by the chromosome number of the sporophyte, which must logically and by definition be regarded as being diploid, or $2n$ only.

PART II: MATERIAL FOR CYTOLOGICAL STUDY

Chromosomes are characteristic of all but the lowest forms of life, therefore all kinds of organisms can be and have been cytologically investigated. Among the higher plants, which are the objects to which the techniques discussed in this book are mainly directed, both herbs and trees have been studied from all parts of the globe, although most chromosome observations have been made in the arctic, boreal and nemoral regions of the northern hemisphere. Chromosome numbers have been reported for plants in a steadily increasing rate since 1882; they have been printed in thousands of publications in numerous languages. Although many species have been repeatedly studied by different cytologists from various parts of their area of distribution, it is our estimate that these reports comprise only a little more than 20% of the probably about 160.000 well-defined biological species of higher plants. These reports have been summarized in numerous chromosome atlases the most recent of which are listed in Appendix II of this text.

Some of the chromosome numbers published for various species, especially by early authors, have later been found to be inexact estimations from difficult material studied with insufficient knowledge or inferior equipment, but still incorrect numbers are being published from various regions. In most cases there is little reason to believe that the authors have made mistakes in counting the chromosomes, but the uncertainty usually derives from inexact taxonomical classification. Botanists all over the world still do not seem to be generally aware of the necessity and importance of exactly and expertly classifying all material that is the basis of a scientific study of whatever kind. In no field of knowledge is this requirement more important than in connection with chromosome numbers because of their general evolutionary importance.

Because of the all important necessity for exact and correct classification of cytological material, a botanist who has gained technical skill in cytological methods, but not become equally familiar with the philosophy and techniques of plant taxonomy, would do right in cooperating with an experienced and critical taxonomist when naming his material. Such a specialist ought preferably to be a botanist who is used to the practice of plant identification by aid of flora manuals, monographs, revisions, original descriptions and herbarium material. However, in most cases an acceptable determination can be made by the collector himself, even when deciding upon subspecific units, although critical cases always ought to be referred to somebody with better facilities or training.

In order to make it possible to send material elsewhere for checking of the determination, or to make it possible for later generations to study the material more closely, it is very important that the collector keeps a pressed herbarium specimen with all pertinent information attached as a voucher. Such vouchers ought to be deposited at a large and permanent official herbarium, and they should be mentioned, preferably by collector's name and number, in connection with the publication of the results. Naturally, a voucher is no guarantee for the correctness of a taxonomical or cytological determination or for the reliability or the skill of the author. Although it is deplorable that most early determinations were made without such a voucher specimen being taken care of, it would be foolish to follow the illogical and naive recommendation of some recent authors, who want to discount the many chromosome counts for which no herbarium specimens are available, since that would be similar to the wisdom of pouring the child out with the bathwater. However, if there is some legitimate reason to doubt the exactness of the information, the availability of a voucher specimen will make it possible for later generations to check and perhaps correct the taxonomical identification, änd so avoid to drop from the literature a report, which is likely to be cytologically correct despite taxonomical faults. The same requirement ought to be made for whatever kind of collections, be they seeds, plant parts treated on the spot, or plants to be transplanted for experimental cultivation for scientific observation by any method of approach. Vouchered or not, however, it is worth repeating that a scrupulous precision in classification of the material and exactness in observation are of an utmost importance in cytology no less than in other sciences.

The technique of collecting and pressing botanical material may be acquired from various books, some of which are listed in Appendix II, or it can be learnt from botanists or amateurs who have practiced plant collection in the classical manner of this old art. The same goes for the later treatment of the herbarium specimens before they are sent to an official herbarium. The herbarium technique can be supplemented by aid of modern photographic methods, which make it possible to preserve even enlarged and exactly colored photographs of details that otherwise might be difficult to study on dry specimens. This is true also for microphotographs of chromosomes and other microscopic details, copies or specimens of which could even be attached to the herbarium vouchers.

CHAPTER 2: MITOTIC MATERIAL

A. General

Although it is often more easy to make good cytological preparations from material cultivated in pots under controlled conditions, fixations of plant tissues made in nature are usually satisfactory. Since temperature, light, and the condition of the soil influence the beginning of and rate of cell division, it is generally recommended that fixation be made early in the day and in cool weather, although we have had considerable success also with collections gathered in the late afternoon of sunny, hot and dry days on the prairies of North America and at the Mediterranean.

Mitotic material is most important for the exact counting and morphological studies of the chromosomes. The best stage of mitosis might seem to be the divisions in the pollen grain or pollen tube, because there only the haploid number need to be observed. Since this division is difficult to locate in plants with which the observer is not familiar, other tissues are more ideal for such observations. We group them below after their suitability or reliability so that we recommend mainly three tissues, though others listed are useful when the first group is out of the question and when there is an urgent need for immediate study that cannot wait for the preparation of the most appropriate material.

B. Recommended tissues

a. Root-tips

The meristematic tissue which is most useful as a source of material with mitotic divisions is the growing root-tip of plants. Cell divisions are very active just above the root-cap but below the region of root-hairs in a zone about one to two millimeters long in the periblem and the plerome. Most of the divisions are longitudinal on the root so the metaphases are frequent in a plane with a transverse section of the tissue. Higher up, where the meristem is mature, endomitoses may occur, especially in certain families, as, e.g. Chenopodiaceae and Aizoaceae but also in others. Then so-called endoploid cells with the double or quadruple chromosome number may occur and perhaps confuse the chromosome number determination, although a skilled cytologist will observe that the chromosomes in such cells often are more or less distinctly paired. Still further from the apex the cells stretch lengthwise while their nuclei remain in a permanent resting stage.

Root-tips may be fixed from wild-growing plants, which then are carefully dug up and cleaned from intruding roots and, preferably, washed. Old and mature roots are usually yellowish or brownish and look dry and shrunk, whereas young and actively growing roots are characterized by being somewhat thicker and brittle and of another color, frequently white, but sometimes bright yellow, green, red, clear brown, or even black. When several such roots have been laid bare and their connection with the plant ascertained, they are broken off 1/4 to 1/2 cm above the tip by aid of tweezers and placed in small vials into which the fixing fluid has been poured to about 1/2 to 1 cm from the bottom.

It is usually easier to obtain good roots from plants which have been transplanted and grown in pots for some time. The most appropriate time of fixation is, then, when the roots have filled the pot all around. The day before the fixation the lump of soil is loosened from the pot by knocking its upper edge cautiously towards a stone or an edge of a bench, when holding the hand over the soil with the base of the plant between two fingers to prevent the lump from falling to pieces, lifting the lump out for a moment and replacing it in the pot in such a way that a thin layer of air remains between the soil and the wall of the pot. The plants are watered properly and left until next morning, when the lump is again lifted from the pot. Then there are usually numerous new roots which have grown into the airspace near the side of the pot during the night, and one can select straight and thick root-tips in the number required for fixation.

b. **Germinating seeds**

Germinating seeds are a very good source of fresh and clean root-tips filled with dividing cells, provided that the roots are not allowed to elongate too much. Seeds can be germinated between wet layers of filter paper in a petri dish and also in clean sand in ordinary daylight, although some species may require more than average room temperature. Some seeds have to be treated with cold or frost for some time before they can sprout, and the germination of many kinds of seeds increases considerably after a few days of treatment with air of high oxygen content. It is important that great care is taken in identifying and pressing the plant from which the seeds were taken, and we also recommend growing plants to maturity from the seeds to secure that the identification has been correct and that no weeds have replaced the original seeds.

c. Cork cambium of trees

Frequently, it is difficult to reach the root-tips or flower-buds of trees, especially large specimens in forests or gardens. Karl Sax (1959; cf. also Hally Sax 1960) worked out a method by which parenchymous growth can be stimulated on tree trunks where it can easily be reached. The bark is carefully cut out into the wood, leaving a hole a couple of centimeters long and broad. The hole must be immediately covered tightly with a sheet of polyethylene to prevent drying. One or two days later, rims of callus tissue can often be observed on the wood. It is filled with cell divisions and can either be treated as other meristematic tissues or, preferably, be squashed directly in a suitable fixative. This method cannot be used with conifers because of their bleeding of resins.

C. Other tissues

a. The shoot apex

The shoot apex or the apex of branches is a reasonable source of mitoses, although all the cells are not cooriented in the same degree as those of the root-tips. The occurrence of chloroplasts and other inclusions in the cell, and also of layers of wax or resins on the outer layers of the shoot or branch, may require special treatments. Since divisions occur only close to the apex, leaves which cover it need to be removed and only a couple of millimeters' long piece of the extreme tip should be cut off and fixed.

b. The leaf tip

Meristematic growth is also met with in the extreme tips of leaves and has been found to be useful when other material has not been available for cytological study. Only tips of very young leaves are worth fixing, because soon all nuclear divisions cease and growth continues through differentiation of individual cells in a permanent resting stage.

c. Corolla in flower-buds

Petals from small buds of various plants have been found to be a reasonable source of mitoses, which then are especially frequent near the margins and low tips.

d. Tillers of grasses

In grasses, active tillers can be stimulated to new growth by cutting back the foliage, but they themselves have also been found to be a reasonable source of meristems. A method to reach the meristem has been worked out by Bennett (1964). With a narrow-bladed scalpel an actively growing tiller is severed from the parental shoot by a downward directed cut between its base and that of the shoot. The outer leaves are carefully removed, until the tender and easily breakable shoot is exposed. An appropriately long piece of the shoot is cut off and immersed in the fixing solution. Details of the method are available in the original publication.

A perhaps simpler method has been recommended by Powell (1968). It is based on locating the youngest internodes and excising an approximately three centimeters long section of the shoot. The lowest cut is made below the last visible node. A longitudinal cut is made through the last node and the conical growing point of the stem with a scalpel, exposing the very young leaves surrounding the growing point to permit contact with the fixing solution.

e. Maturing ovules

Developing ovules are a source of mitotic divisions, and so is also the endosperm of maturing seeds. Both can be easily dissected out and fixed by aid of any of several methods.

f. Staminal hairs

The staminal hairs of *Tradescantia* and other Commelinaceae have long been known as an excellent source for direct observation of cell division and chromosome movements in living cells. They can also be fixed and prepared for studies of chromosome number and form.

g. Other tissues

Various other tissues with active meristems have been used for chromosome studies in many plants, and also meristems stimulated to secondary growth by different methods. The availability of such tissues seems to be limited only by the knowledge and imagination of the research worker himself.

CHAPTER 3: MEIOTIC MATERIAL

If material for studies of the reduction division is required, this can be collected only from flower buds. No general rule as to the size or appearance of such buds can be given, because flowers vary in size as do other parts of the plant, though it may be helpful to observe that in plants with few medium-sized flowers in the inflorescence, buds with meiotic divisions are frequently only one-fifth to one-tenth of the size of the almost ripe buds. In plants with small flowers the buds with divisions tend to be somewhat larger. In grasses meiosis happens before the inflorescence emerges from the surrounding sheath, and in some early flowering shrubs and trees and in plants with bulbs the division may occur during the fall or winter preceding the actual flowering.

In many plants meiosis commences in the afternoon or evening and is completed next morning so the important first division of meiosis could take place in the middle of the night. This requires some enquiry before a large scale and final collection of material is made, to avoid frustration later. In plants with many and small flowers, bulk fixation of the entire inflorescences with flowers at all stages of development is advisible, and selection of appropriate sizes of buds can then be done on the fixed material just before staining or embedding.

PART III: BASIC EQUIPMENT

CHAPTER 1: MICROSCOPE AND ACCESSORIES

Various kinds and qualities of light microscopes can be used for chromosome observation, although such small objects are most easily studied in a good research microscope. We assume that the reader is familiar with their lenses, resolving power, magnification, lighting, mechanical stages, and other important features concerning the handling of microscopes; if not, insufficient knowledge can easily be remedied by aid of special handbooks and assistance from colleagues. Of the normal accessories, special mention needs to be made of the filters, since for successful cytological observation colors complementary to the stain will enhance clarity. A blue filter is appropriate for fuchsin, carmin and orcein, but a yellow-green filter is recommended for gentian or crystal violet.

Equipment for microphotography may simplify the recording of the observations, although frequently it may not give adequate results because chromosomes have a tendency not be be in the same focus plane. We have found that equipment with a Polaroid-Land camera using positive-negative film is far superior to other such outfit, especially because of the possibility to check at once the results on the positive and since the negative is superior for the production of copies for printing.

A good camera lucida is an essential requirement, because without such an equipment an exact counting is possible only for low chromosome numbers. Also, a camera lucida is used for measurements of chromosomes and for making detailed and exact drawings which can include particulars that are observable but not easy to photograph. With the camera lucida should be a stage-micrometer, the scale of which can be drawn for exact measurements of the magnifiction, and an eyepiece micrometer is recommended for direct measurements of observed objects.

Since cytologists work with high magnifications, it is essential that the microscope be furnished with a good mechanical stage. By aid of both the micrometer sides it is possible and indeed essential that the situation of the plates studied can be noted down for later references if needed. However, many cytologists still prefer to mark such places with a ring of India ink on the underside of the slide, because then they are also able to find the place rapidly in another microscope. A still better method for a speedy location of a plate in any microscope is to use the so-called England finder, which is a glass slide marked over the top surface with graticules in such a way, that a reference position can be deduced by direct reading, the relationship between the reference pattern and the locating edges being the same in all finders. The slide is replaced with the finder without altering the position of the fixed stage, and the reference pattern on the finder is recorded as exactly as possible after focusing on the ruled graticule; its position then corresponds to the original position of the observed feature. By reversing the procedure and relocating the graticule on any other microscope, the original slide can easily be placed with the observed feature in exact position.

CHAPTER 2: RECORD BOOK

A small note book, preferably of a size easily carried on excursions, is an absolute necessity for all scientific observations. It is recommended that the same number be used for voucher specimens intended for herbarium or other preservation, individuals to be transplanted for cultivation, seed samples, pictures, and the fixations and preparations themselves. The number ought to be written with a lead pencil because ink may fade or wash away if affected by sunshine or water. It should be written directly on the sheet in which the plant is being pressed, attached with a label to plants to be transplanted so that it can be permanently affixed to them in the pots and in the experimental field, and on a label securely attached to the outside of the vial with the fixation. To prevent mixing of such labels or losing them during transportation or storage, the number may also be written on a small piece of paper which is sunk into the fixing fluid in the vial and allowed to follow the fixation until the number can be transferred to the end of the finished slide.

Various systems of numbering have been invented by botanists of the past, from the very simple to the very complex. Some prefer to use consecutive numbers for each year, distinguishing the year by adding its two last numbers: $1/75$, $\frac{1}{75}$, 1-75, or 75-1. Others have separate series for each family, or for each genus, or for each locality, and some distinguish series by adding letters behind or in front of the numbers, or both. One complicated system uses the two last numbers of the year, two numbers for each month of the year, two numbers for the day of the month, and three or four figures for the collection itself. In that system, 7505230071 would indicate the 71st collection on the 23rd day of May 1975. An impressive and informative number, but hardly practical for simple identification of more detailed information which belongs in the record book itself.

The simpler the system the better. Most experienced herbarium and field botanists, who may make thousands of collections in one summer, mark their collections with consecutive numbers that continue from the first to the last collection throughout their lifetime, perhaps with minor deviations for very special series or for collections made together with others. Such a system has a great advantage and avoids all kinds of complications that could cause some serious error.

Information to be written into the record book may be simple or complex, depending on needs and the wish of the collector, but it is wiser to write too much rather than too little and rely less on memory when time goes by. There ought to be space for the name of the plant, date and locality of collection, what kind of vegetation grew in the same habitat, and other perhaps essential details that might be useful when publishing the results of the study of the material. The name that is given to the plant in the field is likely to be preliminary, but it must be corrected as soon as a final and exact determination has been made, with information about who made the determination, if somebody other than the collector himself is responsible for the identification.

The basic implements for fixing tissues for cytological observation are a strong and a fine pair of tweezers for breaking off root-tips and other tissues, fine scissors, mounted needles, and a scalpel for dissection of flower buds and other material to be fixed. A razor blade is often handy for dividing material into smaller pieces.

Various kinds of vials may be used for the fixation of root-tips and flower buds, made either of glass or of clear and acid-proof plastic. For flower buds, vials that are 50 mm or longer and 20 mm or wider are recommended, whereas for root-tips an appropriate size is 30-40 mm long and 10-15 mm wide, although this depends upon individual taste. Some prefer molded plastic screw-caps to seal the vials, but we have found full satisfaction by using either cork or plastic stoppers.

Ordinarily, the fixing solution does not need to fill more than 5-10 mm of the bottom part of the vial, depending on the size of the material, which must be thoroughly submerged. If the vials are intended for transportation by mail without further preparation, it may be wise to fill them with the solution and secure the stoppers, although fixed material can be drained and mailed in a drop of glycerin, as will be mentioned later.

CHAPTER 4: TOOLS FOR STAINING AND MOUNTING

The basic material for staining and mounting cytological preparations is glass slides of ordinary size and quality, and coverslips, short for squashes and long for sections. It is recommended that the slides be with at least one end etched, on which the appropriate number can be written with a lead pencil. The coverslip should preferably be of a thickness number 0 so that oil immersion can be used. Watchglasses, with or without cover, are good for warming tissues for pretreatment or maceration and for staining. Drop-bottles for stains and bottles for chemicals are best made of acid-proof plastic, and so are also small measuring cylinders and petri dishes. A small thermostat oven is needed for hydrolysis of material intended for squashing, and a spirit lamp is handy for heating slides and smaller objects.

For paraffin embedding a thermostat oven is necessary. Open glass vials or troughs about 20 mm deep and 50 mm wide are recommended for infiltration of paraffin in the oven, and ordinary watchglasses, about 60 mm in diameter, are suggested as being most handy for embedding. For section cutting, a good and exact microtome is a necessity, preferably of a type with a transport band.

For stretching and drying the ribbons when they have been cut and placed in water on the slide, a hot plate or a special slidewarmer, which can be maintained at 45°C, is recommended.

For removing the paraffin and for hydration, staining, and dehydration, various kinds of troughs are available, but we recommend a type with a cradle or sledge in which several slides can be moved simultaneously through the solutions. Mordanting and staining, however, is more easily done in narrower jars with ridges to hold the slides in place, similar jars without ridges for differentiation of the stain, and ridged jars for keeping them in xylene or alcohol until mounting. Special self-holding tweezers are handy tools for the process of staining and differentiation of individual slides.

PART IV: TECHNIQUES OF OBSERVATION

After completing the fixation and staining of appropriate tissues, which thus have become available for microscopical study, the technique of observation remains. When the purpose of the study is the counting of chromosomes, we recommend the use of somatic divisions in the root meristem of both spermatophytes and pteridophytes as the most reliable material, though other tissues may be useful when root-tip material is not easily available.

The simplest technique for counting chromosomes requires a selection, under a comfortably low magnification, of a place on the slide where numerous divisions seem to occur. With an increasingly larger magnification a cell in full metaphase is singled out and put under the highest magnification by aid of immersion oil. With the assistance of a camera lucida, a line is drawn on a paper for every chromosome as in Fig. 26. To securely count their number, each line is crossed. It is recommended that at least three metaphase plates be counted from each of three different roots and, if at all possible, also from at least three different individuals, in order to secure the exact chromosome number and the absence or presence of supernumerary chromosomes or fragments, which occur in some plants. The best plate ought to be selected for photography if possible, and it should be used for exactly drawing the outlines of the chromosomes, to be filled in with India ink as soon as possible (Fig. 26).

Counting the chromosomes from other somatic tissues is similar to that from root-tips, and the number so determined will also be the diploid number, $2n$, typical of the species. When the determination is made on pollen grain mitoses, the number will be n as in meiosis, and if endosperm mitoses are used, the number is usually $3n$.

Some cytologists prefer to determine chromosome numbers from meiotic divisions, and when the number is very high, this may be easier. Then polar view of the metaphases is used, but also polar views of either anaphase. The number counted will be the haploid number, n. Since multivalents and other meiotic irregularities may blurr the picture and disturb the anaphase distribution of the chromosomes, several plates must be counted with considerable care when meiotic material is used.

Fig. 26: Diagram of counting and drawing chromosomes in a camera lucida.

Observation of meiotic divisions is also made for studies of the pairing conditions, chiasma behavior and frequency, occurrence of disturbances due to segmental interchanges, inversions or other structural differences in hybrids, multivalent configurations, and other deviations from the normal condition in hybrids and polyploids. Such observations are usually made on side views of first metaphases or anaphases, although irregularities at other stages may also be of interest.

Cytologists with access to microphotographic equipment prefer to take pictures of the best available plates and configurations, especially when the chromosome numbers are high or when the configurations are complex. Several pictures may be needed to ascertain that chromosomes at different focal planes will not be lost. Some observers even find it desirable to compare the plate and the photograph and strengthen the outlines of the chromosomes in the latter with India ink, especially those chromosomes which may be partially hidden behind others and, thus, at a somewhat different focus. This is most easy when the microphotographic equipment allows the use of a Polaroid-Land camera, but can also be done after traditional microphotography.

When the purpose of the investigation is the study of the karyotype, or of the variation in size and form of the individual chromosomes in the somatic complement, several methods have been invented. Cytologists working with animal and human chromosomes, and also with large plant chromosomes prepared by aid of squash techniques, frequently cut out individual chromosomes from enlarged microphotographs and arrange them in a row in pairs to produce a rough karyogram. Such cut-outs are then placed in the order of size, and individual chromosome pairs are identified by (1) the ratio of the length of the longer arm to the shorter arm, (2) the centromere index, which is the ratio of the length of the shorter arm to the total chromosome length, and (3) the presence or absence of secondary constrictions although they are not always easily detected. Each pair of autosomes is designated by a number beginning from 1 with the longest, and if allosomes occur, they are marked as X or Y. This is an inexact method, though it seems to be sufficient for its purpose.

Exact studies of karyotypes require considerably more sophisticated techniques, beginning with a critical selection of fixation fluids that show minimal influence on the length and morphology of the mitotic chromosomes. The most appropriate and inexpensive fixatives for chromosome morphology in plants are various combinations of the Levitsky fixatives. Since these fixatives usually require sectioning for best results, it follows that for exact karyotype studies the commonly used squash techniques are not recommended, with the exception of the special method proposed by Östergren and Heneen (1962) described later.

When metaphase plates are selected for karyotype study, utmost care is needed so that as many as possible of the chromosomes are in exactly the same plane from one end to the other and also at exactly the same stage of the metaphase, because of variation in their contraction (Bajer 1959). At the largest magnification, each chromosome is drawn separately by aid of the camera lucida, and great efforts are made to mark exactly their ends and the location of the centromere and of possible secondary constrictions. When the chromosome is bent, or the ends and centromere are in different planes, their correct length may be measured by aid of the

95

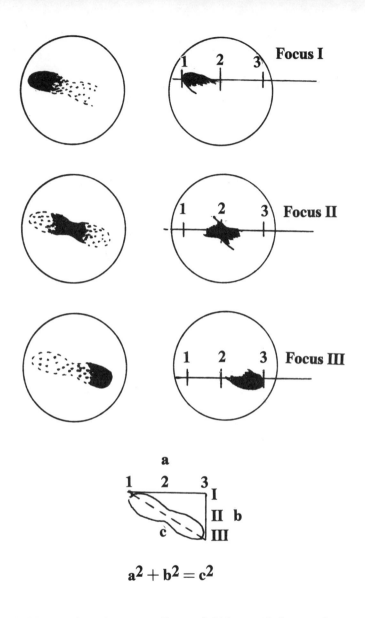

Focus I

Focus II

Focus III

$$a^2 + b^2 = c^2$$

Fig. 27: Measuring a chromosome all parts of which are not in the same plane.

micrometer screw as explained in the manual for each microscope (Fig. 27); the distance measured will, then, be the short sides (a and b) of a right-angled triangle of which the long side (c) can be calculated by aid of Pythagoras' theorem.

Although measurements of the chromosomes may be made by aid of an eyepiece micrometer, more accurate measurements are performed with the scale of a stage-micrometer which is drawn on the paper with the same magnification. Instead of calculating the magnification, the enclosing of such scales with the pictures gives a reliable indication of the enlargement used, and of the actual length of the different chromosomes.

Some students of karyotypes prefer to enlarge the image in the camera lucida considerably by placing the paper at a level much lower than the microscope. However, little is gained from this procedure.

Since variations in measuring the chromosomes are unavoidable (Matérn and Simak 1969; Bentzer and alii 1971), it is recommended that at least ten and preferably twenty or more excellent plates from each individual be selected for every karyotype study. When each chromosome pair has been carefully measured and identified, the average length of each chromosome is plotted in a row, beginning with the longest pair at the left side and representing each pair with a single picture only, except in the case of allosomes or of evident heterozygosity for the length of some of the pairs. The picture so composed is the karyogram in the more exact and original meaning of the term. Since this conventional way of depicting a karyogram ignores the variation observed in the length of individual chromosomes (Bajer 1959), some authors have tried a more exact presentation in a curve form into which the standard deviation is drawn.

It must be left to the individual researcher to select between different methods of making and presenting karyograms on basis of the reasons for which the karyogram is being made and the exactness to be required.

Karyotype studies may be important as a tool for the study of possible relationships of species and genera and of the occurrence of hybridization. They may also be indicative of gradually developing reproductive barriers between related taxa at and below the species level. But the inexactness that is necessarily built into the observations of karyotypes ought to caution against the drawing of too important conclusions from differences or similarities that may be caused by the measuring difficulties mentioned.

CHAPTER 3: SIZE OF POLLEN AND STOMATA

An increase in cell size is a reliable indication of panautoploidy, and it is also frequently useful for identifying hemiautoploids and alloploids as compared to their diploid relatives.

Cell size is most easily studied on pollen grains, the variation of which falls within the normal curve. The pollen is satisfactorily stained in Müntzing's aceto-carmine-glycerin, which is described in Appendix I. The grains are measured either directly in the microscope by aid of an eyepiece micrometer, or from camera lucida drawings considerably enlarged (Fig. 28).

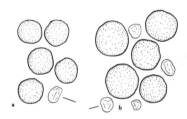

Fig. 28: Pollen size and pollen fertility. a. Diploid. b. Tetraploid.

The size of cells can also be studied from stomatal guard cells (Fig. 29), which are most common on the underside of leaves of most plants. Several methods have been invented to pull small pieces from the lower epidermis of young leaves, the most successful being the dipping of the leaf into hot water for a second or two before cutting and pulling the epidermis with a small scalpel; these pieces are then observed either in aceto-carmine-glycerin or in plain water. Another successful method requires the painting of the leaf with nail polish or plastic (Sinclair and Dunn 1961), which is then pulled off and placed on the slide for a study of the relief of the epidermis.

Since cell size varies within the normal curve, it is essential to measure at least 100 cells from each individual or deme to be compared, and preferably more. Cell size can also vary because of genetical differences

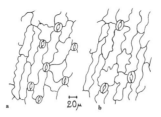

Fig. 29: Size of stomata. a. Diploid. b. Tetraploid. After Pazourková in Hrubý (1961).

between demes at the same level of polyploidy, and it is affected by some environmental factors, so comparison ought to be made only between material grown under similar conditions. When different levels of ploidy are to be compared, the curves are likely to overlap somewhat, because the smallest cells of the higher level are probably smaller than the largest ones of the lower level, and only the biometrically treated averages are to be compared. Many good books in biometrics or biological statistics are available for the beginner and the advanced student, but the recently published volume by Sokal and Rohlf (1969) will satisfy all.

An important characteristic connected with cytological phenomena is the fertility of pollen grains, which is known to vary within and between populations of plants and to reflect the occurrence and frequency of certain meiotic disturbances. It also varies somewhat with certain environmental factors.

Several simple techniques have been invented for the study of pollen fertility, the most easy one being that proposed by Müntzing (1939), who fixed and stained the pollen grains on a slide in a mixture of aceto-carmine and glycerin. A somewhat more sophisticated medium for fixing pollen grains is the aceto-carmine glycerin jelly advocaded by Marks (1954). Since empty pollen tends to stay in the anther when filled pollen grains are shed, it is important that ripe but still undehisced anthers be selected and crushed in the solution on the slide by aid of a pair of tweezers, and the rubble cleaned away before the coverslip is put in place. After a day the filled and round pollen grains have become deeply stained so they can easily be distinguished from the empty and shrunk shells of the abortive grains (Fig. 28), but there is no need for studying the slides so soon, because the glycerin makes them semipermanent. There can be no doubt that the empty pollen is sterile, whereas the method does not separate viable from inviable filled grains. This is also true for seemingly more sophisticated methods using other stains or enzyme tests, which are more time-consuming than they are reliable. The same applies to direct observation of pollen germination under artificial conditions (Löve 1943a), and to a fluorescence method recently recommended by Heslop-Harrison and Heslop-Harrison (1970), since rarely do the results from such experiments warrant their complexity.

Pollen sterility may vary within and between the anthers of the same individual and also between different individuals of the same deme, so a study of 100 or more grains from each preparation is recommended. These variations may be caused by environmental factors, as nourishment or age of the branch, and they have been found to be sizable between flowers at various places on the same fruit tree (Wanscher 1941). Sterility also varies with genetical and cytological differences. There is considerably less diversity in pollen sterility in demes or larger populations of autogamous than of allogamous plants, an observation believed by Müntzing (1939) to be connected with the greater heterozygosity of the latter for various kinds of cytological disturbances.

Studies on the sterility of pollen grains ought to be accompanied by cytological observations of the meiosis, and vice versa (cf. M. Gustafsson 1973). They also need to be followed by biometric comparisons in order to prevent or counteract fallacious conclusions or misunderstandings of the results.

PART V: CYTOTECHNOLOGY

CHAPTER 1: GENERAL REMARKS

Although cell divisions and chromosomes may be observed in living cells in plants and animals, experience has shown that studies of preserved material give results that are superior for numerous reasons. For this purpose, the contents of the cell must be hardened and made insoluble and protected against various kinds of decay and shrinkage. This is done by aid of so-called cytological fixatives, which are solutions or mixtures of selected chemicals, which also will affect the refractive indices of various parts of the cell and thus make them more easily observable after staining with other chemicals. In order to influence certain characteristics of the chromosomes, or to increase the number of some stages of the cell division, pretreatment with some chemicals is often used; it may act as a kind of narcosis. A large number of substances has been used in fixatives, but the most important are certain alcohols, chloroform, formaldehyde, acetic acid, propionic acid, chromium trioxide, picric acid, and osmium tetroxide, and fixing fluids are usually mixtures of two or more of these. We prefer to describe here only those fixatives that are proven to be effective and to produce only a minimum of artifacts, and so can be recommended for general use. However, we refrain to even mention many superfluous fixatives that are recommended by those, whose names are fixed to them, since most of these have in no way been demonstrated to be superior to the simple ones here described. We also abstain from discussing the influence and properties of the components of the mixtures described, because such considerations are available in other books, and they are of little concern to the general cytological practioner for whom this book is written.

In order to make the chromosomes visible in the microscope, the tissues must be either sectioned into a layer one or two cells thick, or spread into a single layer on a slide, and subsequently or simultaneously stained. Sectioning requires that the tissues be treated so that they can be cut into thin sections. This may sometimes be done after freezing, but embedding into paraffin is by far the most superior method, especially for cytological preparations of plants. It is followed by staining according to various procedures, of which we describe the three most useful for this kind of material. Since squashing is considerably faster, though not necessarily a superior method, which avoids the use of a microtome and makes it possible to stain the fixed tissues much more rapidly, it has almost replaced sectioning of material for chromosome study except for more sophisticated investigations. Although different procedures of squashing have been invented by numerous workers, we describe only the basic method which has been proven to be effective for all kinds of material studied in our laboratory, and also recommend only a limited number of stains to be used in connection with squashing. The mode of action of these stains is very different and has been discussed in details by chemists and cytotechnologists, although for our purpose only a reference needs to be made to such discussions.

In order to preserve the stained tissues, they must be sealed under a thin coverslip, either temporarily or permanently. The most useful methods and substances for this are described without discussions of the reason for their selection, since this has also been done in more special texts. However, we recommend the adherance to the chemicals here given rather than experimenting with something new which may or may not cause that the laboriously made slides could be rendered useless for future studies.

CHAPTER 2: SQUASHING TECHNIQUES

A. Tissues of the sporophyte

Although several methods for squashing meristem tissues have been described, they are generally only modifications of a simple technique, which may need to be varied slightly for different stains and tissues. The most convenient material for squash preparations of mitotic chromosomes is the growing root-tip, although other tissues are treated in much the same way.

a. Pretreatment

Tissues are pretreated in various ways in order to facilitate their study. A classical method is based on keeping the material at low but not freezing temperature for 6-24 hours before fixation, either in pots before the root-tips are excised, or as 5-10 mm long cut-off root-tips in water in vials in a refrigerator. This procedure shortens the chromosomes and makes them easier to count, especially in some grasses and other plants with long chromosomes that tend to be crowded in the metaphase plate.

Most pretreatments are, however, made to arrest the divisions at metaphase in order to increase considerably the number of plates available for study. The most effective method for that purpose treats the excised root-tips by placing them in the vial in any one of the following solutions for 2-4 hours:

(i) 0.002 mol. 8-hydroxyquinoline at 10-18°C (cool with running tap water)
(ii) Saturated aqueous solution of a-monobromonapthalene at room temperature (20-25°C)
(iii) 0.01-0.2% aqueous solution of colchicine at room temperature
(iv) Saturated aqueous solution of paradichlorobenzene at room temperature
(v) Saturated aqueous solution of coumarin at room temperature

After any of these pretreatments, the root-tips must be washed thoroughly for a few minutes, for instance by transferring them to another vial with tap water. If a-monobromonaphthalene has been used, washing with 45% acetic acid will help to remove all traces of this oily solution.

The time for pretreatment may need to be varied somewhat for different material. It can be omitted when there is a plenty of root-tips that have grown fast, and when time for the study is limited. It should be avoided if stages later than metaphase are to be observed, because the arrest of metaphases caused by the inactivation of the spindle will result in anaphases to become scanty or absent.

b. **Fixation**

Although various of the standard and sophisticated fixatives described in Appendix I may be used for material to be squashed, results most adequate for studies of the chromosome number are obtained after fixing in the simple acetic-alcohol (1 : 3) fluid attributed to Farmer though already used by some earlier authors, or in the somewhat more complex Carnoy's fixative. The objects are, most conveniently, fixed in the same vial as used for the pretreatment. In case Farmer's fluid is selected, the minimum fixing time is two hours and up to twenty-four hours at room temperature, wheras fixation in Carnoy's fluid may be complete in five minutes, though up to twenty-four hours is recommended for best results. The tissues can be kept in either solution for several months in a refrigerator, but if the fixations are to be stored for any length of time, it is better to replace the fluid with 70% alcohol, in which the objects will keep for months, especially if stored at low temperatures (0-4°C). However, before staining takes place, material stored in alcohol ought to be transferred to 45% acetic acid and kept in it for at least one hour and preferably over night.

We want to emphasize that fixing can be omitted if the tissues are to be stained either in acetic acid stains or according to the Feulgen technique, but a fixation reduces the staining of the cytoplasm and is therefore recommended. Also, for some material, it seems to be advantageous to replace acetic acid with propionic acid, by using Cutter's fluids instead of Farmer's or Carnoy's fixatives. Then, dyes solved in propionic acid are used for staining.

c. Hydrolysis or maceration

When the Feulgen technique is used for staining the tissues, it is necessary to hydrolyse the material. This is done in the same vial as the fixing, replacing the fixative with 10% hydrochloric acid after a rinse for a couple of minutes in warm tap water, and placing the vial in a thermostat oven at 60°C for 10-30 minutes. If an oven is not available, a water-bath can be used. After the hydrolysis, the acid is drained away and the root-tips are rinsed quickly in tap water.

Although hydrolysis is not needed for other staining media, it is recommended prior to carmine or orcein staining because it softens or macerates the root-tips. The same softening will be achieved if the tissues are placed for 5-10 minutes in a mixture of one part concentrated hydrochloric acid and one part 95% alcohol. Also after this, the material should be washed before proceeding further.

d. Staining

For staining in aceto-carmine, aceto-orcein, aceto-lacmoid, or nigrosine, the root-tips are placed in the desired stain for 30 minutes to three hours, either on a covered watchglass or in the vial. If remaining too long in the stain, the cytoplasm tends to become too dark, so care is needed in selecting the optimal staining time for each material. Gentle heating may intensify and quicken the staining.

Objects to be stained according to the Feulgen technique are kept in the vial with leuco-basic fuchsin at room temperature in a dark place (for instance a drawer) for one to three hours, or until the root-tips have become intensely purple.

Root-tips stained in acetic dyes must be squashed immediately, but those stained the Feulgen way can be stored almost indefinitely in 45% acetic acid at temperatures between —14 and —32°C, or they may be kept in water for a few days. Soaking in water before squashing intensifies the stain.

Acetic acid can be replaced by propionic acid in the stains mentioned, and other combinations may be selected from the list of recipes in Appendix I.

e. Squashing

For squashing, an about 2 mm long tip is cut off the root, and its upper part is discarded. The tip is placed on the middle of the slide in an appropriately large drop of solution for squashing.

When carmine, orcein, lacmoid or nigrosine dyes are used, the tip is placed in a drop of 1% solution of the stain and the slide is warmed gently for a couple of seconds over a spirit lamp but never more than so that it does not burn the back of the hand when touching it. Then the slide is placed on the table, preferably on a sheet of white paper to make the tip easily visible, and a coverslip is put on it with the tip under its middle. The tissue will spread somewhat under the gentle pressure from the weight of the coverslip, but it should be flattened by some tapping, for instance with the rubber end of a pencil. The tapping should be continued until the cells have spread into a thin, single layer. If a check under low power of the microscope should reveal that the layer is too thick, more pressure must be added with stronger tapping. A layer of filter paper is put on top of the coverslip and pressed down to suck up any excessive stain. Since the acid is detrimental to the skin, a piece of thin rubber (for instance from a bicycle tube) should be placed on top of the filter paper to prevent direct contact with the fingers, or a rubber glove worn for protection. It is important that all side movements of the coverslip be avoided.

If the Feulgen method is employed, the technique is the same, except that the root-tip is covered either with a drop of 45% acetic acid or, preferably, with a drop of 1% aceto-carmine, which intensifies the staining.

The slides can be either temporarily sealed or mounted permanently by aid of any of several methods described on a later page.

f. Rapid techniques

A much simplified procedure that can be recommended for chromosome studies from root-tips when time is limited, has been described by Levan (1972). The roots are fixed for 5-10 minutes in Levan's fluid (see Appendix I), which is a mixture of glacial acetic acid, hydrochloric acid, and water. They are then macerated in the fixative by cautious heating for some few minutes. The root-tips are transferred to a slide with a drop of 2% orcein in 60% acetic acid and chopped with a needle. After the cells have taken good stain, the coverslip is added and the material squashed in the way described above. Although the method is recommended for root-tips, it is likely to be useful also for other tissues rich in somatic divisions. For those with a few divisions only, a pretreatment in 0.002 mol. 8-hydroxyquinoline for a few hours before fixation may enhance the results.

A perhaps still quicker method is recommended by Marks (1973). Fresh or pretreated root-tips are fixed and hydrolysed in 5-normal HCl for 15 minutes at room temperature. After washing, the root-tips are macerated on a slide with a needle in a drop of 0.05% solution of toluidine blue (G.T. Gurr's 04800) made up in the McIlvaine buffer, which is a mixture of 0.1 mol. citric acid and 0.2 mol. Na_2HPO_4 buffering at pH 4.0. The stain becomes rapidly dark blue. It is not necessary to heat the slide before adding a coverslip and applying gentle finger pressure through a double thickness of filter paper covered with a piece of rubber to protect the fingers. Seal for temporary use or mount permanently in euparal.

g. **Double staining**

Differential staining of the chromosomes and the nucleolus can be added to preparations which have been stained according to the Feulgen technique, by aid of a method originally described by Semmens and Bhaduri (1941). For this, the coverslip is removed in acetic-alcohol (1 : 1) after squashing, and the slide with the squashed material is transferred to 80% alcohol through two five minute changes. Mordant for 45-60 minutes in a clear and saturated solution of sodium carbonate in 80% alcohol. Three five minute changes in 70% alcohol follow before the slide is brought to 90% alcohol and stained for 10-15 minutes in a specially prepared solution of light green as described in Appendix I. Differentiation is in 80% alcohol with a trace of sodium carbonate, dehydration by changes through 90% and 95% to absolute alcohol, after which the preparation is cleared in xylene-alcohol grades for mounting in Canada balsam. The nucleolus will be light green and stand sharply differentiated against the dark red chromosomes.

Another method for double staining in a single step, which leaves the chromosomes brownish red and the nucleolus and cytoplasm green, was recommended by Kurnick and Ris (1948). The root-tips are stained in a mixture of aceto-orcein and fast green for a few minutes (cf. Appendix I for recipe) and squashed as usually.

h. A technique for chromosome morphology

A modification for obtaining material acceptable for the study of chromosome morphology in squash preparations has been worked out by Östergren and Heneen (1962). The root-tips are pretreated with 0.002 mol. 8-hydroxyquinoline for 3-4 hours at 15°C before being fixed overnight in Östergren and Heneen's fluid (see recipes in Appendix I). Hydrolyse for 8 minutes in 10% hydrochloric acid at 60°C, and stain according to the Feulgen technique for two hours before squashing as usually.

B. Tissues of the gametophyte

a. Pollen mother cells

When possible, the size of the flower buds at which meiosis takes place is determined before any large scale fixation is commenced. An inflorescence is selected and a bud about 1/5 to 1/7 of the size of a ripe bud is picked out for removal of one of its anthers, which is placed on a slide with a drop of an acetic or propionic stain. After a coverslip has been placed on the top of the anther, the slide is gently heated over a flame to about 50°C. The coverslip is tapped and then pressed firmly under a piece of filter paper and held steadily to prevent sideways movements of the coverslip. The slide is examined under a low power magnification to determine the stage of gametogenesis. If pollen grains or tetrads are already met with, the slide is discarded and the lowest bud in the inflorescence studied in the same way. Should even that bud include tetrads, another smaller inflorescence must be picked, but if the lowest bud is at a pre-meiotic stage, the correct size of bud is likely to be in the middle of the inflorescence.

In many cases, direct fixation and staining may be used to make meiotic preparations for final study. A few anthers from the same flower are, then, placed on the slide, fragmented with a needle or a pair of tweezers in a drop of the selected stain, and the debris of anther wall and the tapetum removed before the coverslip is placed on the slide, whereafter it is heated gently and the coverslip pressed down as described in connection with squashing mitotic tissues. However, such preparations tend to be unevenly stained, and their contents also unevenly spread, so other methods, which seem to be more elaborate at first glance, could actually prove to be more effective and time-saving.

It is recommended that the size of the buds with meiotic divisions be determined prior to entering on a large scale fixation, but since this is not always possible when fixing in nature, a wide range of inflorescences with buds reaching to about one-fifth to one-tenth of the length of buds ripe for flowering can be selected for fixation. They are placed in Farmer's acetic-alcohol for five minutes or more. If the fixed material needs to be kept long before staining, it is preferably transferred to 70% alcohol within 24 hours from fixation and kept in a refrigerator, and then transferred again to 45% acetic acid for at least an hour before staining is attempted.

When the material is to be stained, each vial is emptied into a petri dish and appropriate sizes of buds selected by aid of a rapid check following the method described above. Anthers are removed and placed in warm water for rinsing before being transferred to a vial with 10% hydrochloric acid in which they are hydrolysed at 60°C for five to ten minutes. In the case of small flowers, or when for some reason it might be worth while waiting with the selection of anther size until the squashing takes place, the entire inflorescences may be kept untouched until the staining is completed.

After the hydrolysis has been done, the anthers are rinsed quickly in water. Then they can be stained in the vial, either in carmine, orcein or other acetic or propionic stains for one to four hours, or according to the Feulgen technique for two to three hours in darkness.

For squashing, one or more anthers are placed on the slide in a drop of the stain or in 45% acetic acid. After a gentle heating of the slide, the coverslip is put in place, the anthers crushed by tapping, and the pollen mother cells spread thinly with a light pressing as described earlier for root-tips. At this stage, the slide may be studied, but if it is to be stored even for only a few hours, the edge of the coverslip needs to be sealed with any of the temporary sealants mentioned in Appendix I. It can then be kept for several months, especially if stored in a refrigerator.

If for some reason the slide is to be made permanent, this can be done by aid of any of several methods described on a later page. The most simple way requires that the coverslip be removed and the slide transferred through solutions appropriate to the mounting medium selected. This can be made easier if the slides are thinly smeared before use with Mayer's albumen and carefully dried over a flame for a few seconds, because then most of the material is likely to stick to the slide when the coverslip is detached.

b. Embryo-sac mother cells

Ovules or pieces of the ovary, which have been dissected free of the surrounding tissues, may be fixed in Carnoy's fluid for one day, hardened in 95% alcohol for another day or two, and gradually hydrated through a decreasing series of alcohol. Hydrolysing and staining in acetic-lacmoid or according to the Feulgen method is recommended. A high number of embryo-sac mother cells may have to be fixed before appropriate meiotic stages have been found in an amount needed for a proper analysis of the meiotic divisions.

e. Mature embryo-sacs

The ovaries are fixed in Carnoy's fluid for at least two days when material of mature embryo-sacs is sought. The fixative is then replaced with a 4% aqueous solution of iron alum with iron acetate (add ten drops of a saturated solution of iron acetate in 45% acetic acid to 10 cm³ of 4% iron alum). The vials are placed in a waterbath at 75°C for three minutes. The solution is drained and the tissue washed with two changes of warm water and one change of cold water for 2-3 minutes each. Maceration in 50% hydrochloric acid for 10 minutes. Rinse in several changes of water for at least 20 minutes.

For staining, the ovaries are placed on a slide in a drop of ferric aceto-carmine or aceto-orcein into which the ovules are scraped from the placentae. The debris is removed, and the ovules are tapped with a

114

scalpel until the cells are separated. A coverslip is applied and the slide is heated gently before pressing the coverslip for squashing.

Although it is possible to get some notion of the development of the embryo by aid of this method, it cannot replace standard approaches which require sectioning the ovules by a microtome.

d. Endosperm

In order to study the chromosomes in the endosperm, the fertilized ovules are pretreated in 0.002 mol. 8-hydroxyquinoline for three hours or with a saturated aqueous solution of a-monobromonaphthalene for two hours. The tissues are rinsed in tap water and fixed in Farmer's or Carnoy's fluid. They are hydrated in 80%, 70%, 50% 30% alcohol and in water before being hydrolysed in 10% hydrochloric acid at 60°C for 8-10 minutes. Staining can be either in ferric-aceto-carmine or in aceto-orcein for one to several hours or by the Feulgen technique for two hours, before the endosperm is dissected out and squashed in 45% acetic acid, or in the stain as in the case of mitotic cells.

CHAPTER 3: SECTIONING TECHNIQUES

A. Introduction

Although the rapid squash techniques have largely replaced the older and more elaborate sectioning techniques for many kinds of cytological observations, the classical methods are still required for handling certain tissues and also for work of greater exactness when for various reasons the fast methods do not give satisfactory results. This is so, for instance, for most embryological studies, in which it is important that the relation of each tissue to another is not disrupted.It is also done when very exact karyotypic studies require the use of sophisticated fixations after which ordinary squashes cannot readily be made. Sectioning is preferred by those, who make mass fixations during expeditions or travels away from laboratory facilities, because it is considerably more difficult to prepare such material for successful squashing than to treat it by aid of the microtome and the methods of complicated staining which always have been known to give excellent results.

B. Fixation, dehydration and infiltration

Tissues to be used for sectioning can be fixed in any fixing fluid available. When fixations are made in nature with chromosome counting in mind, the simplest and most reliable fluids for all kinds of tissues and all kinds of plants are variations of the Navashin's fixative, of which we prefer the Svalöv modification although all the other variants seem to be equally effective. However, for studies of karyotypes and chromosome morphology, variations of the Levitsky's fixative are recommended. These simple fluids are considerably less expensive than some recently invented and complicated solutions, and at least almost as effective in most cases.

Fixations made in Navashin type solutions can be kept in the vials indefinitely, but they must not be transferred to 70% alcohol until after 24 hours when the fluid has turned clear green. At the time of transfer, the tissues are washed quickly in warm tap water in the vials, but if soil particles adhere to the root-tips, these are rolled on a moist piece of cloth or filter paper for cleaning. When placing the objects in the vials with 70% alcohol for one hour or indefinite time, it is advisable to add also a small piece of paper, e.g. 15 × 5 mm large, on which the number of the fixation has been written in pencil. This paper will then follow the material up to the paraffin cake. Also, when the root-tips or other tissues are small and colorless, it can simplify their handling if some drops of eosin are added to the 70% alcohol to color them red.

Since paraffin is immiscible with water and aqueous solutions, the tissues must be dehydrated before they can be embedded. For dehydration several agents that mix with alcohol and with paraffin have been used in the past, but those which permit the fastest procedure at the same time as they avoid shrinkage and hardening of the tissue and are not too poisonous if inhaled, are n-butyl alcohol or tertiary butyl alcohol, either of which we recommend, because in our experience they are safer and more reliable than chloroform or dioxan often used by others. The following schedule for dehydration is appropriate, giving the safe minimum time of each fluid: 70% alcohol for one hour, 95% alcohol for one hour but not more than two hours, absolute alcohol for one hour but not more than two hours, absolute alcohol: butyl alcohol (2 : 1) for half an hour, absolute alcohol: butyl alcohol (1 : 2) for half an hour, butyl alcohol for at least half an hour. These liquids can be reused many times, so when the changes are made, this is done by pouring each mixture carefully out of the vial, slanting the stopper in the opening to prevent the tissues to flow out with the fluid which is poured into a vessel or a dish from which accidentally lost objects may be easily recovered. The fluid should be filtered back into the bottle through a piece of gauze or linen to prevent contamination.

When the tissues have been in pure butyl alcohol for half an hour, the entire content of the vial is poured into small dishes, preferably 2 cm deep and 4-5 cm wide, into which melted tissuemat or paraffin with a melting point of 53-54°C has been filled to about 1 cm from the bottom and allowed to solidify. These dishes are placed in an oven at 55-56°C which is just sufficient to keep the paraffin melted (avoid higher temperatures to prevent hardening of the tissues), in which they are kept for replacing the butyl alcohol in the tissues with paraffin, a process that takes 12-24 hours.

C. Embedding

When the smell of butyl alcohol has vanished, usually after 12-24 hours in the thermostat which ought to be ventilated as the room in which it is kept, the infiltration of the paraffin is completed and embedding can take place. Instead of the still much used paper trays of earlier generations, we recommend the use of flat-bottomed watchglasses, 5-6 cm wide, which during embedding can be placed on the top of a 4-5 cm wide dish of the type used for the paraffin infiltration. The watchglasses must be carefully cleaned, first in soapy water and then in alcohol, to avoid any film of fat, and they must be evenly but thinly smeared inside with a mixture of 60% alcohol and glycerine immediately before they are to be used, in order to prevent the paraffin to adhere to the glass. The hot dish with the root-tips or other tissues is removed from the oven, the label with the number lifted out with a pair of tweezers which have been heated over a spirit flame, and placed at the side. Then the dish is carefully rotated to get the tissues in motion, and the entire contents are emptied into the watchglass with a quick tilt. If the objects are not in motion, they may remain in the dish when the paraffin is poured out: they can then be transferred to the watchglass with the warm pair of tweezers or a warm mounted needle.

The tissues are quickly orientated and grouped in the paraffin by aid of needles or tweezers which are repeatedly heated over a spirit flame to prevent the paraffin to solidify on them. If root-tips are involved, they are placed in groups of four to six with the tips towards the outside and exactly in line and all the roots as parallel as possible. To hold them in place, they are gently pressed into the thin layer of solid paraffin that immediately forms at the bottom of the watchglass. A thin skin is allowed to form on the surface of the paraffin before the flattened label is attached to the middle of the cake with heated tweezers. Then the watchglass is floated carefully on cold water in a bowl to harden the paraffin at the bottom and top sides of the cake, and finally gently submerged in the water. Sudden cooling is desirable because it makes the paraffin crystals smaller than slow cooling does. It is recommended to slip the watchglass into the cold water slightly obliquely to prevent the still liquid paraffin in the middle of the cake to burst through the congealed film on the top, sucking with it the embedded objects. When the paraffin has solidified, the cakes will float up by themselves if the glycerin film has been complete; otherwise they can be removed from the glass by aid of a gentle push with a finger or a knife. The cakes can be arranged on a filter paper on a table away from heat for drying, and they can be kept in a cool place until sectioned. In that condition, the paraffin cakes and the tissues may be kept indefinitely at room temperature, protected from dust and direct sunshine.

D. Sectioning

For sectioning cytological material a good microtome is needed, preferably one with a transport band onto which the ribbon can move automatically. We have found special microtomes from Leitz and Reicherts to be widely superior to other makes for this purpose.

When the paraffin cake from which the sectioning should be made has been selected, the number on its label is transferred to the end of the slide to be used. With a sharp scalpel or a razor blade the piece of the cake immediately below the selected group of tissues is carefully levelled. Then a cube with the tissues in its middle is cut out of the paraffin. It has to be fastened to the chuck or paraffin holder of the microtome by aid of paraffin that is melted to the holder with a scalpel. The cube is thrust into the melted paraffin with the tips of the roots facing away from the holder, and the paraffin then melted around it to anchor it properly. The holder with the embedded specimens may be dipped into cold water or allowed to cool in the air for 10-15 minutes; it is recommended that several holders be in use at the same time to avoid delays during the time when the paraffin is cooling down.

When the paraffin cube on the holder has again become properly cold, the sides of the block are trimmed to as parallel a rectangular square as possible with a razor blade with the tissues close to one of the longer sides, though avoiding to cut into or too close to the specimens and retaining the broader base of attachment intact. The holder is attached to the microtome with the tissue side up and adjusted in such a way that the knife will cut out the material in a 90° angle seen both from the side and from above. If the material to be cut is root-tips of plants with unknown chromosome size, we have found that sectioning at 12 micrometers is generally acceptable, though plants with larger chromosomes may need to be sectioned thicker and those with smaller chromosomes thinner. Each section will remain on the knife until the next pushes it forward, fusing it with edge to edge to form a ribbon, which preferably will move with the transport band but could also be led away from the knife by aid of a moist camel hair brush held in the left hand, in case the microtome does not have a transport band. The ribbons of sections are carefully cut into lengths somewhat shorter than the coverslip to be used, either on the transport band or on a clean sheet of paper.

The slide with the number on is coated with a thin layer of Mayer's albumen and covered with cold water by aid of a small camel hair brush; the same moist brush is used to transfer the cut ribbons onto the slide and to arrange them in close rows, with the shiny (lower) face down. When the ribbons are not parallel, they can be further straightened and stretched with two needles if the slide is placed on a warm plate (40-50°C) for a few minutes. When the ribbons have been arranged and stretched, the slide is tipped sideways to let most of the water carefully drain off. The slide is placed to dry, preferably for 24 hours in an oven about 35°C warm, or on a cool part of the hotplate, or in a dustfree cupboard at room temperature.

Several difficulties are sometimes encountered in section cutting. They may be caused by static electricity in too dry rooms, by too low or too high room temperature, by too soft or too hard paraffin, by wrong tilt of the knife, by inexactness in trimming the paraffin blocks, by dirt or smear on the back of the knife, by nicks in the knife, by dullness of the knife, by insufficient removing of alcohol from the paraffin, and by several other inaccuracies in the process followed or in the equipment itself. Some experimenting usually solves these problems.

E. **Removing the paraffin**

The paraffin has to be removed from the material so the sections can be moved down to water for staining. For this the slides, preferably several at the same time, are placed in a specially made cradle and kept in a trough or slide-jar with xylene overnight or for at least a couple of hours and then transferred for five minutes to another trough with clean xylene. If for some reason such a waiting must be avoided, the paraffin can be removed almost instantly by melting it carefully over a flame and dipping the slide into cold xylene, but also by dipping it into xylene which has been heated (in a waterbath) to about 50-55°C. When the paraffin has been dissolved, the cradle with the slides is dipped a few times into. troughs each with the following liquids: xylene: absolute alcohol (1 : 1), absolute alcohol, 95% alcohol, 70% alcohol, 50% alcohol, 20% alcohol, and tap water. Bleaching of the cytoplasm is done after the 95% alcohol in a freshly made solution of 30 cm^3 of H_2O_2 (30%) and 70 cm^3 of 95% alcohol overnight or at least for a few hours; this is required for objects fixed in solutions with osmic acid and recommended for most mitotic tissues after any fixation, although this stage may often be avoided after fixation in the Svalöv modification of Navashin's fluid. When the staining is to be in crystal violet, a few drops of acetic acid in the water will improve the result. If the staining of some slides has to wait overnight for some unforeseen reason, it is necessary to move the cradle back to 70% alcohol to prevent damaging the tissues by too long soaking in water.

F. Staining

Several methods for staining chromosomes after sectioning have been in use during the past 90-odd years, although only three of these are generally employed at present. These are the crystal violet or gentian violet method, the Heidenhain's haematoxylin schedule, and the Feulgen technique, all of which give excellent results by aid of relatively simple procedures.

a. The crystal violet method

Several versions of this simple and rapid method have been described. We favor the following, which uses Gram's aqueous solution of iodine and potassium iodide for mordanting, and mixes aniline blue and crystal violet for staining. The recipes for the solutions are listed in Appendix I.

The exact time for staining and differentiation must be worked out for each material, because it depends on the nature of the tissue, the thickness of the sections, and the fixing solutions used. Therefore, the times given here are only tentative, although they are close to being correct for general root-tip material fixed in the Svalöv modification of Navashin's fluid and sectioned 12 micrometers thick.

The slides are moved from water through the following schedule:

1) Crystal violet with anilin blue, 5-30 minutes

2) Rinse in tap water with some drops of acetic acid

3) Mordant in Gram's aqueous iodine solution, 3-60 seconds

4) Rinse in 95% alcohol, 10-20 seconds or until color begins to run off

5) Differentiate in absolute alcohol as desired by dipping for 1-3 minutes

6) Cleanse in very thin terpineol by dipping a few times or for up to 1-2 minutes. This step can be omitted if the highest possible accuracy for staining is not required

7) Arrest the differentiation by rapidly dipping the slide into xylene; examine under the microscope

8) A short dip in eugenol or clove oil, or differentiate further if needed

9) Three changes of xylene, about 10 minutes each

10) Mount in a neutral medium, as, e.g., neutral Canada balsam or permount.

If the differentiation in the alcohols and eugenol has been insufficient so that the cytoplasm remains heavily or unevenly stained, the slide may be moved backwards from stage 8 to 95% alcohol for further differentiation.

If the chromosomes are too weakly stained, the slides are taken through the alcohol series to water and mordanted after the water stage either in 1% chromic acid for 12-24 hours, or in a saturated aqueous solution of picric acid for 15 minutes to 2 hours, or in a saturated aqueous solution of picric acid mixed (1 : 1) with glacial acetic acid for 15 minutes to 2 hours. Rinse in tap water and repeat the staining process as above.

If the contrast between the chromosomes and cytoplasm is poor, a better differentiation can often be achieved by dipping the slides into ammoniac-alcohol (5 drops of ammoniac in 100 cm^3 of 95% alcohol) between steps 4 and 5 above.

If some crystals are formed on the slides after staining, they can be prevented by rinsing before and after the crystal violet in 5% acetic acid.

b. Heidenhain's haematoxylin

This classical method gives extraordinary results and the stain remains permanent for decades. Most schedules are rather elaborate and slow, but we recommend the following relatively fast procedure:

1) Mordant in 4% aqueous ferriammoniumsulphate for 5 minutes to 24 hours

2) Rinse in running tap water for 10 minutes and in distilled water for 10-20 seconds

3) Stain in haematoxylin for 30 minutes to 24 hours

4) Rinse in running tap water for 10 minutes and in distilled water for 10-20 seconds

5) Differentiate in 2% aqueous ferriammoniumsulphate and check result in the microscope

6) Arrest the differentiation in water; rinse in water for 30 minutes

7) Transfer to xylene through the alcohol and alcohol-xylene series.

If the mordant and the staining solution are kept at 60° C, the mordanting can be reduced to one minute and the staining to half an hour.

The method is better for mitoses than for meiotic divisions, but the cytoplasm in thick sections tends to remain somewhat stained. The schedule is especially recommended for exact karyotype studies, particularly after fixation in Levitsky's fluids.

c. The Feulgen method

The Feulgen schedule for staining chromosomes after sectioning gives a transparent but fairly permanent stain, which does not need differentiation, although it requires a rather exact timing of hydrolysis. The results may also be affected by insufficient reduction of osmic acid by bleaching after fixation in certain solutions, by staining of the cytoplasm if fixatives containing formalin are not thoroughly rinsed out, by temperature and other environmental variations, and by a poor kind of fuchsin. The following procedure is recommended for root-tip sections:

1) Hydrolyse in 10% hydrochloric acid at 60°C. The time for hydrolysis is 4-8 minutes after fixation in Farmer's, Carnoy's or other fluids without chromic acid, and about 15 minutes (6-30) after fixation in fluids containing chromic acid. If the material has been stored in alcohol, the time for hydrolysis may have to be extended, and sections of tissues from some plant families require a longer time of hydrolysis, probably because of some chemical interference.

2) Rinse rapidly in water and stain in leuco-basic fuchsin for 30 minutes to 3 hours. This is done in darkness or in a well shaded jar.

3) Rinse in running tap water for 15 minutes. The sections must not come into contact with the air, so avoid splashing. If organelles other than the chromosomes tend to become stained, rinse three times in SO_2 water for 10 minutes each time and once more in distilled water.

4) Dip slides through a series of 20%, 40%, 70%, 95%, and absolute alcohol and mount in euparal or other alcohol-soluble medium, or transfer further through alcohol-xylene and mount in neutral Canada balsam or permount.

For tissues that have been fixed in solutions without acetic acid or in aqueous fixatives with acetic acid, the results of a Feulgen stain may be improved by treating the slides with 1% chromic acid for 4-6 hours after the hydration series, rinsing thoroughly in water, and keeping them in 70% alcohol for 4 hours prior to hydrolysis.

The staining can also be improved if the slides are heated in 60°C warm water for 5 minutes before hydrolysis.

The staining is likewise improved if the trough with the slides is cooled rapidly by immersion in cold water after hydrolysis and the slides transferred to cold 10% hydrochloric acid for a few seconds prior to rinsing in water and staining.

d. **Double staining**

Differential staining of the chromosomes and the nucleolus can be made on sectioned material by following a modification, by Levan, of the method by Semmens and Bhaduri (1941) described in connection with the squashing technique. After Feulgen staining, the slides are rinsed in tap water for 15 minutes, and then transferred relatively quickly through distilled water, 30%, 50%, and 80% alcohol to one hour's mordanting in a saturated solution of sodium carbonate in 80% alcohol. Staining in light green for 15 minutes. Differentiate in a solution of 40 cm³ of 80% alcohol mixed with 100 cm³ of saturated sodium carbonate in 80% alcohol. Rinse in 80% alcohol for 15 minutes, and transfer to absolute alcohol and to xylene as usually for mounting.

CHAPTER 4: SEALING AND MOUNTING

A. Introduction

Slides prepared with the squash method can be studied under the microscope without additional treatment, because the coverslip is tightly stuck to the glass through the stain itself. Although they can be kept for a few days, especially if stored in a refrigerator, they tend to spoil through evaporation if nothing is done to prevent this. Therefore, they ought to be sealed or mounted as soon as the need to keep them has been established. Slides prepared with the sectioning method inevitably must be permanently mounted even for observation, but they can be kept infinitely in xylene.

B. Temporary sealing

Sealing squash preparations for temporary storage can be done in various ways so that evaporation of the medium is delayed or prevented for a limited period of time, but these methods should not be used if a more premanent storage is necessary. Drying of the preparation can be delayed if lactic acid is mixed into the staining fluid, for instance by replacing aceto-carmine or aceto-orcein with lacto-propiono-orcein. Destruction of the material can also be slowed down by mixing glycerin into the staining media, as is preferably done in studies of pollen fertility. The most effective and simplest method of making such slides semipermanent is, however, to seal the edges of the coverslip with rubber cement, which can be easily removed if a decision is made to mount the slide permanently, or with paraffin, paraffin-mastic, gum, gum-mastic, aceto-gelatin, poly-ethylene-glycol, quick-drying cellulose paint, gold size, nail polish or other media which can be applied around the edges with a brush or melted around them with an L-shaped metal rod. The preparation of some such media is described in Appendix I.

C. Permanent mounting

When the sectioning technique has been used, the last step in completing the slides is to mount a coverslip permanently on top of the stained sections with some resin which usually is dissolved in xylene or alcohol. Permanent mounts of squashes, however, require some other techniques, although even for these the only preservative media which are sound from all points of view are of the resin type. For chromosome studies the coverslip must be at least the number 0 quality or even thinner to allow the use of immersion oil for observation in the microscope.

It is necessary that the mounting medium has a refractive index between or close to that of the glass slide and that of the tissue so light passing through will not be lost. The medium should be stable and not change or decompose with age, it should be sufficiently dilute to drop readily off a glass rod, and it must harden quickly and prevent destaining through changes in acidity. In general, Canada balsam dissolved in xylene or alcohol has all these properties, although it is slightly acid, a characteristic which can be somewhat counteracted by keeping a piece of marble or limestone in the bottle in which it is stored. Another widely used medium is a resin mixture called euparal which is soluble both in xylene and alcohol, and a third recommended is permount, dissolved in oil of turpentine and therefore soluble also in xylene. If the mounting medium is dissolved in alcohol, the tissue can be mounted directly from absolute alcohol without a transfer through a xylene series; this is often to be recommended after Feulgen staining, whereas the xylene soluble media used after crystal violet staining also serve to stop further differentiation. Other less used mounting media are diaphane, lacto-phenol, clarite X, polyvinyl alcohol, and several oils, but none of these are as easy to use or of the same general quality for mounting cytological material as are Canada balsam, euparol, and permount.

Several techniques have been described to make squash preparations permanent. Generally, they belong to three categories, depending upon if they require the removal of the coverslip or not, or if the mounting medium is to be included in the solution in which the tissues are squashed.

One-step staining-mounting methods for squashes would be the ideal solution for the squash technique, but most of those so far proposed have not been widely used mainly because they thicken the substratum and so make the spreading of the cells difficult. Relatively successful methods of this kind are those invented by Zirkle (1940), using gelatin and sorbitol, and venetian turpentine, phenol and propionic acid, respectively, in his two schedules; by Traub (1953a, b), whose methods utilize arabinic acid and sorbitol; and by Marks (1954) who bases his approach on glycerin jelly. An improved procedure of Zirkle's methods was proposed by Haunold (1968), who followed the Feulgen technique in staining the root-tips and squashed them in propiono-carmine with a drop of venetian turpentine. Numerous other proposals for a one-step permanent squash technique have been made, but since it is our experience that every one of them is more cumbersome than they are practical and hardly ever result in preparations that deserve to be called durable, we do not recommend their application and refrain from describing any of them closely. Those interested in trying these methods or improving them in some way or another, however, can find ample references in the text by Darlington and LaCour (1962), and Sharma and Sharma (1965), and in the original papers by those who have already tried them. Several of these methods have been published in Stain Technology.

Methods for making slides permanent after squashing without removing the coverslip are all based on the same principle. The tissues are dehydrated and mounted by applying certain solutions around the edges of the coverslip so they can be sucked in and so replace the original fluid. If the coverslip has been temporarily sealed, the sealant must first be scraped and cleaned away carefully with an appropriate solvent. The slide with the untouched coverslip is placed in a staining trough, or in a small petri dish with two broken wooden matches (head removed!) on which the ends of the slide can rest, covered with or inverted in 45% acetic acid, and left until the coverslip loosens or falls off. If the slide has been treated with Mayer's albumen before the squashing was made, the material will most likely stay on it, but otherwise it may at least partially remain on the coverslip. The slide and/or the coverslip with the squashed tissue is carefully transferred to 70% alcohol for a couple of minutes, followed by two equally long rinses in absolute alcohol from which the slide can be mounted in euparal. The slide is then dried in an oven or on a warm plate. If mounting in Canada balsam dissolved in xylene is preferred, the alcohol treatment must be followed by a series of alcohol: xylene (1 : 1), alcohol: xylene (1 : 2), and pure xylene, five to fifteen minutes each, before thin Canada balsam is applied.

Squash preparations can also be permanently mounted by aid of the so-called pincer method recommended by Nicoloff and Daskaloff (1964). After squashing, the coverslip is fixed by metal pinchers to the slide and passed successively through a series of 50% alcohol for 15 minutes, 95% alcohol for 30 minutes, absolute alcohol for 60 minutes, after which the slide with the coverslip and pinchers can be placed in thin euparal or thin Canada balsam dissolved in alcohol for 12 hours before being rapidly rinsed in alcohol and dried. The slide can also be transferred further through absolute alcohol: xylene (1 : 1) for 2 hours, xylene for 6-8 hours, to thin Canada balsam dissolved in xylene in which it then stays for 3-12 hours. Then it is rinsed rapidly in xylene to remove the balsam from the outside of the slide, the pinchers are removed, and the slide is allowed to dry overnight or longer, either at room temperature or in a thermostat oven.

If permanent mounts are required, it is least time-consuming and most safe to make the squashes on clean slides which have been thinly smeared with Mayer's albumen and dried gently over a flame. When the squash is made on slides so treated, the tissues will stick to the albumen layer in much the same way as do sectioned ribbons. If the coverslip has been pretreated with silicon (e.g. Dri-film 9987 from General Electric), it will drop off easily without harming the tissue when the slide is inverted into the staining trough with 45% acetic acid. When the coverslip has fallen off, the slide can be transferred successively through a series of acetic acid: absolute alcohol in the proportions 1:1, 1:3, and 1:9, to absolute alcohol, to be mounted in euparal or in Canada balsam dissolved in alcohol, or further through an alcohol-xylene series to xylene for mounting in Canada balsam or another resin dissolvable in xylene. The slide is left to dry at room temperature or in an oven overnight.

Another much recommended approach to the removal of the coverslip is the quick-freeze method originally described by Conger and Fairchild (1953), using dry ice or solid carbon dioxide, but later modified by others for the use of liquid air. After squashing and preliminary observation, the slide is frozen for half a minute or more. The coverslip is lifted by inserting a blade of a knife and the slide is immediately and before thawing placed in a bath of absolute alcohol in which it stays for five minutes. It is then transferred to a second bath of absolute alcohol for two or three minutes and either mounted directly in euparal or Canada balsam dissolved in alcohol, or moved further through an alcohol-xylene series to xylene and mounted in Canada balsam dissolved in xylene. Thereafter the slide is left to dry overnight.

129

CHAPTER 5: MISCELLANEOUS PROCEDURES

A. **Restaining old slides**

Old slides tend to fade for various reasons, but they can be restained. For removing the coverslip, the slide is placed in a jar with xylene, if the original mounting medium was so dissolved, or in 95% alcohol for alcohol-based mounting media. It may take several days for the coverslip to fall off, but this can be hastened by heating the solvent to about 50°C, preferably by keeping the jar in a thermostat oven but also by placing it in a waterbath on an electrical plate. If the original stain was acetic, the slide must be transferred through an alcohol series to 45% acetic acid and to a fresh acetic stain. If the slide had been stained in crystal violet or related stains, it must be transferred through an alcohol-xylene series to alcohol and then down to water after which it can be restained in the original way. But if the Feulgen method had been used, the preparation cannot be rehydrolysed and must be restained in an acetic stain if originally fixed in acetic fixatives, in crystal violet if originally fixed in aqueous fixatives, and in haematoxylin if the original fixative was alcoholic. Preparations which were originally stained in crystal violet can be hydrolysed and restained according to the Feulgen method, especially if the tissues had been fixed in a Flemming type fixative, although the results may not always be as satisfactory as when crystal violet is again used for the restaining.

B. **Dried-out fixations**

It is the frustrating experience of many cytologists, who have spent tin.e fixing material for later study when travelling or even at home, to find that some of the vials contain only dry objects, because the stopper has been loose or some other accident has allowed the fixing fluid to evaporate. If the object was fixed in solutions with acetic acid and alcohol, nothing can be done, but the tissues can be rescued if the fluid contained chromic acid or osmic acid, which leave a crust of residues.

A dried-out tissue should be washed in warm water for a few hours to rinse away the crust on its outside, or the crust can be rinsed away with strong sulphuric acid, after which the material is left in either 10% alcohol or 10% formalin for a few days for rehydration. Then it is transferred to 70% alcohol and treated as if it had never dried-out.

C. **Dry preservation**

In addition to accidents as those just mentioned, botanists on expeditions to far-away places, and also those who need to send fixed material by mail, have often felt the disadvantage of having to bring considerable amount of acids and then having to keep the tissues for a long time in more or less dangerous chemical solutions. The former problem can be partly avoided by using Langlet's fluid for the fixation. The latter problem has been studied by some authors who have intentionally dried fixations for a longer or shorter time. We have tried some such methods recommended by others, and have found the following modification of a proposal originally made by Vaarama (1950) to be most useful in such situations.

The tissues (root-tips, flower-buds, etc.) can be fixed in any kind of fixative. After they have been thoroughly fixed for more than the minimum time required for the kind of fluid selected, they are washed and stored in 70% alcohol for at least one hour and preferably somewhat longer. Then most of the alcohol is drained from the vial and a drop of glycerin added. A few minutes later, the tissues are poured out onto a piece of filter paper, which is folded and allowed to dry at room temperature. Naturally, the number of the collection has been written on the paper, which is then put into a small envelope together with all pertinent information. In this condition, the objects can be transported without risk and stored for several months and probably years.

When the material is to be processed further, it is taken out of the paper and rehydrated in the vials with 10% alcohol for several days, then transferred to 50% alcohol for two hours and to 70% alcohol for at least one hour. Due to the fragility of the chromosomes fixed in alcoholic solutions, such fixations are sometimes difficult to squash after dry preservation without fragmenting or shattering the chromosomes, but they can easily be embedded in paraffin and sectioned and stained with crystal violet or with the Feulgen technique.

D. Storage of pollen

It is sometimes desirable to hybridize races or species that flower at different times, but methods to force flowering at the same period are often difficult and always require considerable and expensive facilities. However, some live pollen can be stored. If needed on a large scale, pollen from ripe anthers is stored in petri dishes or bottles which can be placed in desiccators with humidity at 50% maintained by glycerin in water, and kept at 0-8°C. On a smaller scale, pollen can be stored in vials, on the bottom of which there is a 1 cm thick layer of calcium chloride covered with a loose-fitting plug of cotton-wool. The mature anthers are placed in a small container made from cut-off corners of a thin paperbag or an envelope and placed above the cotton. The vial is corked, properly labelled, and placed upright in a refrigerator at 0-8°C.

In this way, pollen of plants with binucleate pollen may maintain its germination capacity for one to several months, whereas trinucleate pollen tends to lose this capacity in a few days or at least few weeks time, despite desiccation and cold storage.

E. Recycling coverslips

Coverslips, especially those very thin qualities in long sizes used for permanent section preparations, are expensive and ought to be recycled when the slide can be discarded. They may be separated from the slide by immersing in the solvent of the mounting medium for some days, preferably in a thermostat oven or in a waterbath 50-60°C warm. The slides themselves are much less valuable than the coverslips and can be discarded. except when cheap labor is available for their cleaning. The coverslips are washed in a thin and warm soda solution, then transferred to a warm soap solution, clean water, and at last through 70% alcohol to xylene, in which they can be stored until used, if it is not found to be practicable to let them dry individually spread on and covered by filter paper.

If the coverslips are to be used at once for mounting in the same kind of medium as previously used, the procedure can be made much quicker. The new slide with its sections is taken from the xylene or alcohol and a small drop of the thin mounting medium added. The old slide is heated over a spirit flame, the coverslip pushed sideways with a finger or a pair of tweezers and transferred directly to the new slide. No cleaning is necessary.

F. Mixing alcohol

When preparing the alcohol series for dehydration or rehydration, 96% alcohol should be used as a base because it is much less expensive than is absolute alcohol. Sufficient exactness in the strength of the mixtures will be reached by using a cylinder marked for, e.g., 100 cm^3, pouring 96% alcohol up to the 70 cm^3 mark if 70% strength is required, and filling the cylinder with water, not necessarily distilled, up to the 96 cm^3 mark. The procedure is repeated with alcohol up to the proper level for other strengths.

In laboratories which are equipped for physiological or chemical research, the high cost of commercial absolute alcohol can be reduced by distilling 96% alcohol by aid of the following method: Mix two liters of 96% alcohol with 200 g of calcium chloride. Reflex for six hours in a four liters flask. Cool overnight. Add 25 g of calcium chloride, shake well, and distil at up to 80° C in a Kjeldahl apparatus.

G. Etching slides

Slides with at least one end etched are preferred for cytological work, because it is then easy to write numbers and other information on the etched part with a lead pencil and carry it through the solutions. If such slides or funds to purchase them are not available, a short application of diamond ink (fluoric acid) with a stiff brush will easily change an inexpensive slide into one with one or even two ends beautifully etched. Wash the slide at once in water to arrest the etching, and protect yourself and other objects from splashing by the dangerous acid.

PART VI: PUBLISHING THE RESULTS

Those, who feel qualified to advise on how scientific results ought to be made available to others, have written scholarly volumes on this subject. We refer the reader to these books, with the unpretending remark that he should feel free to ignore the advice as to how such information should be written and what style utilized, because the main requirement is that it should be done in an interesting and clear manner.

A short and catching title is better than a long one giving the exact contents of the paper. Chromosome informations may need to be supplemented by simple line drawings and statistical or biometrical information on comparative studies of various morphological dissimilarities, if such have been observed. Many good books introduce the beginner to the basics of statistics and some help him to advance in the subject, whereas still others are both introductory and advanced, as for instance the recent comprehensive text by Sokal and Rohlf (1969). But instructive books in basic drawing for biologists are almost unknown so the short text by Bethke (1969) stands alone in its excellence.

Even apparently simple studies by beginners in the field of chromosome study may result in observations which are new to science so that it would be a sin to let them remain unpublished. In the realm of plant chromosome numbers, many species still are cytologically unknown, so that counting the chromosome number for a plant for which it has not been previously reported is a definite and valuable addition to knowledge. By consulting some of the chromosome atlases listed in Appendix II, it is possible to find out if the observation is new or not. If it is new, it is not necessary to write a special paper to make this known, except if additional observations should warrant this. Instead, the result can be speedily published by sending it, with appropriate information about the origin of the material and of the voucher number and the place where the voucher is kept, to the editor of the chromosome number reports, which are published in every number of the international journal Taxon.

APPENDIX I: SOME STANDARD RECIPES

CHAPTER 1: FIXING SOLUTIONS

A. Alcoholic fixatives

a. Farmer's fluid

Glacial acetic acid. . . . . 1 part
Absolute ethyl alcohol . . . . 3 parts

This is the oldest universal fixative for cytological material. It is effective for all kinds of tissues, plant, animal and fungal. Mix just before use or at least daily. Fix for 2 and up to 24 hours at room temperature. The material can be stored indefinitely in the fluid if kept at 0°C or lower temperatures, but if this cannot be done, transfer to 70% alcohol for storage, and to 45% acetic acid for at least an hour before squashing in acetic stain after such storage. The fluid is sometimes modified by mixing the ingredients in the proportions 1:1, 1:2, or 3:2, but the advantage of this is doubtful.

b. Carnoy's fluid

Glacial acetic acid. . . . . 1 part
Chloroform . . . . 3 parts
Absolute ethyl alcohol . . . . 6 parts

A general fixative used in the same way as Farmer's fluid, but it is often preferred for fixation of flower buds because it causes less hardening of the tissues. Mix just before use or daily. Fixing time 5 minutes to 24 hours; store in refrigerator or transfer to 70% alcohol and then to 45% acetic acid for at least an hour before squashing in acetic stains. Among modifications of the proportions are 1:3:4, 1:1:3, and 1:1:1 which have been recommended for certain plant families, though the advantage is doubtful.

For preparations to be stained in Belling's ferric aceto-carmine some find it advantageous to add a trace of iron acetate to the fixing fluid.

c. Cutter's fluids

1. Original Cutter

Propionic acid. . . . . 1 part
95% ethyl alcohol . . . . 3 parts

This modification of acetic-alcohol is recommended when staining in solutions of propionic acid is planned. Used and treated as Farmer's fluid.

2. Hyde and Gardella's modification

Propionic acid. . . . . $100 cm^3$
95% ethyl alcohol . . . . $100 cm^3$
Ferric hydroxide . . . . $0.4 g$

This modification is recommended for plants with small chromosomes.

3. Newcomer's modification

Propionic acid. . . . . 1 part
Chloroform. . . . . 1 part
Absolute ethyl alcohol . . . . 2 parts

This modification is recommended for flower buds or for replacement of Carnoy's fluid when staining in propionic acid solutions.

Fixing in any of these solutions is for 12 to 24 hours. Transfer to 70% alcohol for storage and then to 45% propionic acid for at least an hour before staining and squashing in dyes solved in propionic acid. All these solutions are to be mixed immediately before fixing, or at least for the day.

d. Newcomer's fluid

Isopropyl alcohol . . . . 6 parts
Propionic acid. . . . . 3 parts
Petroleum ether . . . . 1 part
Acetone. . . . . 1 part
Dioxane . . . . 1 part

This is a universal and stable fixative for plant and animal tissues for cytological squashes or sections. Mix just before use. Fix for 12 to 24 hours. The fixations can be kept indefinitely in the fixative in a refrigerator, or must be transferred to 70% alcohol for storage. After such storage, transfer to 45% propionic acid for at least an hour before staining and squashing in dyes solved in propionic acid.

B. Formalin fixatives

a. Navashin's fluid

Solution A:

Chromium trioxide	1.5 g
Glacial acetic acid	10 cm³
Distilled water.	90 cm³

Solution B:

40% formaldehyde	40 cm³
Distilled water.	60 cm³

The two stock solutions are mixed in equal proportions immediately before being used. Minimum fixing time is 24 hours, but the tissues can be kept in the fluid indefinitely, although rinsing in tap water after at least 24 hours or after the solution has turned green, and storing in 70% alcohol, is recommended. This is a suitable fixative for root-tips and flower buds, but for the latter a short (1-2 minutes) prefixation in Farmer's fluid or in 96% ethyl alcohol is recommended, especially for waxy buds. In this and other fixatives, distilled water may be replaced with tap water without detrimental effects.

b. The Svalöv modification

Solution A:

Chromium trioxide	1 g
Glacial acetic acid	10 cm³
Distilled water	85 cm³

Solution B:

40% formaldehyde.................................... 30 cm³
95% ethyl alcohol .... 10 cm³
Distilled water .... 55 cm³

Mixed and used as the original Navashin's fluid.

c. Randolph's modification

Solution A:

Chromium trioxide .... 1 g
Glacial acetic acid .... 7 cm³
Distilled water .... 93 cm³

Solution B:

40% formaldehyde.................................... 30 cm³
Distilled water .... 70 cm³

Mixed and used as the original Navashin's fluid.

d. Hill and Myer's modification

Solution A:

Chromium trioxide .... 1 g
Propionic acid .... 15 cm³
Distilled water .... 85 cm³

Solution B:

40% formaldehyde.................................... 30 cm³
95% ethyl alcohol.................................... 10 cm³
Distilled water .... 60 cm³

Mixed and used as the original Navashin's fluid.

e. Langlet's modification

Solution A:

Chromium trioxide	1 g
Glacial acetic acid	10 cm³
Distilled water	8 cm³

Solution B:

40% formaldehyde	30 cm³
95% ethyl alcohol	10 cm³
Distilled water	130 cm³

Mix A and B in the proportions 9A:1B just before fixation. Use as the original Navashin's fluid.

This modification is recommended for travels when the equipment has to be kept to a minimum. Langlet (1946) recommends the use of vials prefilled with 1.8 cm³ of solution B into which 0.2 cm³ of solution A, or 6-7 drops from a dropper, are added immediately before fixation. The tissues can be kept in the solution indefinitely, or transferred to 70% alcohol after at least 24 hours in the fixative.

f. Levitsky's fluids

Solution A:

Chromium trioxide	1 g
Distilled water	100 cm³

Solution B:

40% formaldehyde	25 cm³
Distilled water	75 cm³

Mixed in different proportions, such as 4A:1B, 3A:2B, 1A:1B, etc., this is a good general fixative and excellent for karyotype studies. Mix immediately before use, fix for 24 hours or longer, but for extended storage transfer to 70% alcohol is recommended.

g. **Bouin-Allen's fluid**

Saturated aqueous picric acid .... 75 cm³
40% formaldehyde....................................... 25 cm³
Glacial acetic acid .... 5 cm³
Chromium trioxide .... 1.5 g
Urea .... 2 g

The stock solution of the three first ingredients is heated to 37-39°C before using, and the chromium trioxide and urea added under stirring. For studies of *Oenothera* meiosis, Cleland (1972) added only 1 g of chromium trioxide and substituted 1 g of maltose or lactose for the urea. Fixing period is 4-12 hours after which the tissues ought to be transferred to 70% alcohol, which is changed repeatedly until no more yellow color is extracted. This is a good general fixative for flower buds.

h. **Hagerup's fluid**

Solution A:

Glacial acetic acid .... 20 cm³
Absolute ethyl alcohol................................... 60 cm³
Chloroform.. 1 cm³

Solution B:

Chromium trioxide .... 2 g
Tap water .... 100 cm³

Solution C:

40% formaldehyde

Prefix with solution A for 15 minutes; drain and reuse the solution when mixing in the proportions 5A:5B:1C. Fix for 4 to 24 hours, store in 70% alcohol. This is a good general fixative for flower buds and superior for sporophylls of *Equisetum* and cones and other thick inflorescences of various spermatophytes.

C. Osmic acid fixatives

a. Flemming's fluids

1. Strong Flemming's fluid

Chromic acid (1%)	75 cm³
Glacial acetic acid	5 cm³
Osmic acid (2%)	20 cm³

2. Medium Flemming's fluid

Chromic acid (1%)	30 cm³
Acetic acid (5%)	25 cm³
Osmic acid (2%)	10 cm³

3. Weak Flemming's fluid

Chromic acid (1%)	25 cm³
Acetic acid (1%)	10 cm³
Osmic acid (2%)	5 cm³
Distilled water	55 cm³

The solutions are mixed immediately before use. Fixation time is 1 to 24 hours. Rinse in tap water for several hours or overnight, and store in 70% alcohol. Strong Flemming's fluid is used for bulk fixation of flower buds or root-tips, but the other two solutions are recommended for sensitive embryological and meiotic material to be sectioned.

The chromic acid and osmic acid are made from chromium trioxide and osmium tetroxide mixed into distilled water.

b. LaCour's fluids

	2BD	2BE	2BX
Chromic acid (2%)	100 cm³	100 cm³	100 cm³
Potassium bichromate (2%)	100 cm³	100 cm³	100 cm³
Osmic acid (2%)	60 cm³	32 cm³	120 cm³
Acetic acid (10%)	30 cm³	12 cm³	60 cm³
Saponin (1%)	20 cm³	10 cm³	10 cm³
Distilled water	210 cm³	90 cm³	50 cm³

The ingredients are mixed from stock solutions immediately before use.

2BD is a good general fixative, 2BE is recommended for plant material, and 2BX is for bulk fixation. Fixing time for all is 24 hours, followed by rinsing in running water for 3 to 12 hours, and transfer to 70% alcohol for storage.

D. Other fixatives

a. Östergren and Heneen's fluid

Methanol	60 cm³
Chloroform	30 cm³
Distilled water	20 cm³
Picric acid	1 g
2,4-dinitrophenol	1 g
Mercuric chloride	1 g

Mix the three fluids and dissolve the chemicals by adding one at a time, remembering the fact that dry picric acid is a strong explosive. Keep the solution in a tightly closed bottle. Fix for 12 to 24 hours or for a few days. Transfer the tissues directly from the fixative to warm hydrochloric acid for hydrolysis for 8 minutes at 60°C for staining according to the Feulgen technique.

b. Levan's fluid

Glacial acetic acid . . . . 6 parts
Water . . . . 3 parts
1-normal hydrochloric acid . . . . 1 part

Mix immediately before use. Fixing is for some minutes to an hour.

CHAPTER 2: SOME COMMON STAINS

A. Acetic and propionic stains

For preparation of a 2% stock solution of carmine, orcein or lacmoid, add 2 g of any of these chemicals to a mixture of 55 cm^3 of distilled water and 45 cm^3 of glacial acetic acid which has been cautiously heated to boiling. Continue to boil gently until the stain is dissolved. Cool down and filter. Store in corked bottles.

Since 1% solution is most frequently used for staining and squashing, dilute the 2% solution with equal amounts of 45% acetic acid before use.

If nigrosine is to be used, a 4% stock solution is made by adding 4 g of alcohol soluble nigrosine powder to 100 cm^3 of 45% acetic acid, which is boiled gently for 5 to 10 minutes. Cool and filter and keep in a tightly corked bottle for at least ten days. Dilute to 1% before use.

Belling's (1926) ferric aceto-carmine is 1% aceto-carmine to which has been added a trace or a few drops of ferric acetate. It is recommended for meiotic divisions for squashing.

Propionic acid may be used instead of acetic acid and the procedure of preparation of the solutions is similar. When lacto-propiono-orcein is preferred, a 2% solution is prepared by dissolving 2 g of orcein in a mixture of 23 cm^3 of propionic acid and 23 cm^3 of lactic acid at room temperature. Make up to 100 cm^3 by adding distilled water, shake well, and filter. This stain is recommended for root-tips and flower buds after fixation in Cutter's fluids; the slides will keep somewhat longer without sealing because of the lactic acid.

Synthetic orceins have been found to be weaker than natural orceins, so we recommend the use of the latter dyes only, preferably from the British Drughouses or from European continental sources. If American synthetic dyes are used, the amount has to be increased, sometimes considerably, to reach the same strength.

B. Other stains

a. Crystal violet

Dissolve 1 g of crystal violet in 100 cm³ of distilled water with constant stirring and boiling. If the stain is certified, filtration is not needed, but otherwise it is recommended. Cool before using, and if possible allow to stand for a week and mature. Add distilled water to make a weaker solution if necessary. The staining ability of the solution increases with age and usage up to a certain degree, so a complete change of solution is rarely needed.

b. Crystal violet with anilin

Dissolve 2.5 g of crystal violet in 12 cm³ of 96% alcohol, and, separately, 2 cm³ of aniline in 98 cm³ of distilled water. Mix the solutions and stir until even. This mixture, undiluted or further diluted with distilled water, gives a deeper and more uniform staining of somatic divisions than pure crystal violet, and with less staining of the cytoplasm.

c. Heidenhain's haematoxylin

Dissolve 0.5 g of haematoxylin crystals in a mixture of 10 cm³ of 96% alcohol and 90 cm³ of distilled water. Filter and store in darkness in a corked bottle for 6 to 8 weeks before use.

d. Feulgen's leuco-basic fuchsin

To prepare a 0.5% solution, dissolve 1 g of certified pure basic fuchsin by pouring gradually over it 200 cm³ of boiling distilled water. Shake well and cool to 50°C. Add 20 cm³ of concentrated hydrochloric acid. Cool down to room temperature before adding 2 g of potassium meta-bisulphite. Pour the solution into a bottle wrapped in dark paper and store in a dark and cool chamber. If the solution is transparent and straw-colored after 24 hours, it is ready for use, but if otherwise colored, add 1 g of activated charcoal powder, shake well and keep overnight before filtering quickly.

The solution may be kept for a long time in a tightly sealed bottle in a dark and cool place. Occasionally the contents become colored, but this can be remedied by adding a little SO_2 water and shaking until the content is clear. The same solution may be reused almost endlessly if kept away from light and cleared occasionally as mentioned.

A prepared Feulgen solution available from various drug or chemical houses is recommended for time-saving and for securing high quality material.

Sometimes the Feulgen reaction fails, probably because of the chemical composition of the plants involved. The method is still useful if the leuco-basic fuchsin is replaced with other basic dyes of the triphenyl methan group. The most effective substitute seems to be basic magenta, a 1% solution of which is made by dissolving 1 g of the dye in 100 cm³ of boiling water. Decolorize with 2 g sodium hydrosulphite and 15 cm³ of 1-normal hydrochloric acid and add 85 cm³ of water to make an 0.5% solution. If the fluid is still colored, add activated charcoal powder and filter (Golecha 1968).

e. Light green

Dissolve 0.5 g of light green SF and 0.5 g of picric acid in 100 cm³ of 90% alcohol. Bring to boiling, filter, and store in a corked bottle.

f. Fast green

For preparing a mixture of fast green and aceto-carmine for double staining in one operation, three stock solutions have to be made as the first step: (1) Dissolve 1 g of fast green FCF in 100 cm³ of absolute alcohol to make a 1% solution; (2) Prepare a 2 mol. solution of NaCl by dissolving 117 g of table salt (iodine-free) in one liter of water; (3) Prepare a 1% aceto-carmine solution. To produce the stain, mix 3 cm³ of (1) and 2 cm³ of (2) with 27 cm³ of (3). Store in a tight bottle in a cool place and use as ordinary aceto-carmine for squashing.

C. Other solutions

a. 8-hydroxyquinoline

For pretreatment of excised root-tips, a 0.002 mol. solution is prepared by dissolving 0.58 g of 8-hydroxyquinoline in 200 cm³ of distilled water (Tjio and Levan 1950).

b. Coumarin, a-monobromonaphthalene, and paradichlorobenzene

To prepare a saturated solution of any of these chemicals for pretreatment of excised root-tips, place 5 g of their crystals in a bottle filled with distilled water. Shake well, and replenish the crystals so that some will always remain unsolved on the bottom of the bottle.

c. Colchicine

To prepare an 0.5% solution, make a paste from 0.5 g of colchicine powder with absolute alcohol. Continue to add alcohol dropwise until all the colchicine is dissolved. Add very slowly distilled water to this solution to make up to 100 cm³. If the water is added too quickly, or if the solution is poured into water, the colchicine will crystallize. Keep this stock solution in a tight bottle and mix to 0.1 or 0.2% solution when needed for pretreatment, by slowly adding appropriate amounts of water.

The same stock solution may be used for preparation of weak colchicine for treatment of seeds and growing stems of buds of plants in order to produce polyploid tissues, but for that purpose a solution should be only 0.001 to 0.0001% (cf. Lawrence 1968).

d. Gram's iodine

Dissolve 1 g of pure iodine and 2 g of pure potassium iodide in 200 cm³ of distilled water. Shake to hasten dissolution, but heating is unnecessary.

e. SO₂ water

Mix 5 cm³ of 1-normal hydrochloric acid and 5 cm³ of 10% potassium metabisulphite with 100 cm³ of distilled water. Shake well, and store in a closed bottle.

CHAPTER 3: SEALING AND MOUNTING MEDIA

A. Semipermanent sealing

a. Paraffin-mastic

Weigh equal amounts of mastic gum powder and paraffin. Heat the gum gently in a bowl until melted, then add slowly pieces of paraffin and stir until the mixture is homogeneous. Pour into a small petri dish for cooling and keeping. Use to seal around the coverslip with an L-shaped metal rod heated over a spirit flame. If necessary to remove the sealant from the slide, the paraffin-mastic can be dissolved in xylene.

b. Aceto-gelatin

Dissolve 10 g of gelatin powder in 100 cm³ of 50% acetic acid. Allow the gelatin to swell, then heat carefully to dissolve and mix it thoroughly. For sealing around coverslips, apply with a thin glass rod. Sealing may be removed by immersing the slide in 45% acetic acid.

c. Polyethylene-glycol

Weigh equal amounts of polyethylene-glycol 4000 and mastic gum powder. Heat the polyethylene-glycol in a porcelain dish and add the mastic gum gradually while stirring with a glass rod. Apply as paraffin-mastic.

d. Gold size

Melt 24 parts of linseed oil with 1 part of red lead, and 1/3 part of umber. Boil gently for three hours. Pour off the clear liquid and mix with equal parts of white lead and yellow ochre and boil again for some minutes. Pour off the clear liquid and keep in a closed bottle. Apply around the edges of the coverslip with a point of a thin brush; dry at room temperature.

B. Permanent mounting

a. Canada balsam

The most generally used mounting medium is Canada balsam dissolved in xylene, with which it can be diluted. Some prefer Canada balsam dissolved in alcohol. It is made from dry Canada balsam powder which is mixed into benzyl alcohol under gentle heating for dissolving, and further diluted with absolute alcohol. The bottle with Canada balsam of either kind should be kept in a dark place to avoid darkening, and a piece of marble or limestone in the bottle will keep it neutral.

b. Euparal

This medium has a higher refractive index than Canada balsam, but it intensifies some stains. If cloudiness appears during mounting, gentle heating will help. Euparal can be diluted with butyl alcohol, and it is also soluble in xylene and ethyl alcohol.

c. Permount

This resin has about the same refractive index as Canada balsam, and it is essentially similar to it in its effects on various stains. It is dissolved in oil of turpentine and can be diluted with xylene.

CHAPTER 4: MISCELLANEOUS

A. Mayer's albumen

Separate the yolk from the white of a fresh egg and mix with an equal volume of neutral glycerin, the latter serving to prevent desiccation. Then dissolve 1 g of sodium salicylate or aspirin in a minimum of water and add to the mixture as a preservative. When well mixed, filter through a cloth or in a small Buchner funnel in vacuum.

Dried egg white may replace the fresh egg white. Then 5 g of dry albumen is mixed with 100 cm^3 of 0.5% NaCl solution and shaken carefully at intervals for a day, avoiding frothing. Filter to recover about 90-95% of the mixture as a clear filtrate, which is mixed evenly with glyerin. Add a solution of a little aspirin in a trace of water to avoid fouling.

B. Müntzing's aceto-carmine-glycerine

Mix equal amounts of 1% aceto-carmine and neutral glycerin. The mixture is mainly used for the staining and semipermanent preservation of pollen grains for fertility investigations.

C. Immersion oil

Immersion oil for microscopy at high magnification is made by mixing 152 g of liquid paraffin and 48 g of a-monobromonaphthalene. The refractive index of the mixture is 1.515, and it is not necessary to wipe it immediately off the objective or the slide.

APPENDIX II: BIBLIOGRAPHY

A. Current chromosomes atlases

BOLKHOVSKIKH, Z., GRIF, T., MATVEYEVA, T. and ZAKARYEVA, O. (1969) - Chromosome numbers of flowering plants. - Leningrad.

CHIARUGI, A. (1960) - Tavola cromosomiche delle Pteridophyta. - Caryologia 13: 27-150.

DARLINGTON, C.D. and WYLIE, A.P. (1955) - Chromosome atlas of cultivated plants. - London.

FABBRI, F. (1963) - Primo supplemento alle "Tavole cromosomiche delle Pteridophyta" di Alberto Chiarugi. - Caryologia 16: 237-335.

FABBRI, F. (1965) - Secondo supplemento alle "Tavole cromosomiche delle Pteridophyta" di Alberto Chiarugi. - Caryologia 18: 675-731.

LÖVE, Á. and LÖVE, D. (1961) - Chromosome numbers of Central and Northwest European plant species. - Opera Botanica 5: 1-581.

LÖVE, Á. and LÖVE, D. (1974) - Cytotaxonomical atlas of the Slovenian flora. - Lehre.

LÖVE, Á. and LÖVE, D. (1975) - Cytotaxonomical atlas of the arctic flora. - Lehre.

MOORE, R.J. (ed.) (1973) - Index to plant chromosome numbers 1967-1971. - Regnum Vegetabile 90: 1-539.

B. Some cytogenetics and evolution texts consulted

ANDERSON, E. (1949) - Introgressive hybridization. - New York.

BARTHELMESS, A. (1952) - Vererbungswissenschaft. - Freiburg und München.

BRESLAVETS, L.P. (1963) - Poliploidiya v prirode i opyte. - Moskva.

CLELAND, R.E. (1972) - Oenothera cytogenetics and evolution. - London and New York.

DARLINGTON, C.D. (1937) - Recent advances in cytology. 2nd edition. - London.

DARLINGTON, C.D. (1958) - Evolution of genetic systems. Revised 2nd edition. - Edinburgh.

DARLINGTON, C.D. (1963) - Chromosome botany and the origins of cultivated plants. Revised 2nd edition. - London.

DARLINGTON, C.D. (1965) - Cytology. - London.

DARWIN, C. (1859) - The origin of species. - London.

DAWSON, G.W.P. (1962) - An introduction to the cytogenetics of polyploids. - Oxford.

DOBZHANSKY, T. (1951) - Genetics and the origin of species. 3rd edition, revised. - New York.

DOBZHANSKY, T. (1970) - Genetics and the evolutionary process. - New York and London.

GRANT, V. (1971) - Plant speciation. - New York and London.

GRANT, V. (1975) - Genetics of flowering plants. - New York and London.

GÜNTHER, E. (1969) - Grundriss der Genetik. - Jena.

GUSTAFSSON, Å. (1946-1947) - Apomixis in higher plants. - Acta Univ. Lund. N.S. II, 42,3, 43,2, 12: 1-370.

HEILBRONN, A. und KOSSWIG, C. (1966) - Principia genetica. Grunderkenntnisse und Grundbegriffe der Vererbungswissenschaft. 2. neubearbeitete Auflage. - Hamburg und Berlin.

HRUBÝ, K. (1961) - Genetika. - Praha.

JOHN, B. and LEWIS, K.R. (1965) - The meiotic system. - Wien and New York.

JOHN, B. and LEWIS, K.R. (1968) - The chromosome complement. - Wien and New York.

JOHN, B. and LEWIS, K.R. (1969) - The chromosome cycle. - Wien and New York.

KHUSH, G.S. (1973) - Cytogenetics of aneuploids. - New York and London.

KIHLMAN, B.A. (1966) - Actions of chemicals on dividing cells. - Englewood Cliffs.

LAWRENCE, W.J.C. (1968) - Plant breeding. - London and New York.

MAHESWARI, P. (1950) - An introduction to the embryology of angiosperms. - London.

MANTON, I. (1950) - Problems of cytology and evolution in the Pteridophyta. - Cambridge.

MARGULIS, L. (1970) - Origin of eukaryotic cells. - New Haven and London.

MAYR, E. (1942) - Systematics and the origin of species. - New York.

MAYR, E. (1963) - Animal species and evolution. - Cambridge, Mass.

McLEISH, J. and SNOAD, B. (1958) - Looking at chromosome. - London.

MÜNTZING, A. (1967) - Genetics, basic and applied. - Stockholm.

MÜNTZING, A. (1971) - Ärftlighetsforskning. 4 upplagan. - Stockholm.

RIEGER, R. (1963) - Die Genommutationen. - Jena.

RIEGER, R. und MICHAELIS, A. (1958) - Genetisches und cytogenetisches Wörterbuch. 2. Auflage. - Berlin, Göttingen, Heidelberg.

RIEGER, R. und MICHAELIS, A. (1967) - Chromosomenmutationen. - Jena.

RIEGER, R., MICHAELIS, A. and GREEN, M.M. (1968) - A glossary of genetics and cytogenetics, classical and molecular. - Berlin, Heidelberg, New York.

RUTISHAUSER, A. (1967) - Fortpflanzungsmodus and Meiose apomiktischer Blüten-pflanzen. - Wien and New York.

SHARP, L.W. (1934) - Introduction to cytology. Third edition. - New York and London.

SHARP, L.W. (1943) - Fundamentals of cytology. - New York and London.

STEBBINS, G.L. (1950) - Variation and evolution in plants. - New York.

STEBBINS, G.L. (1971) - Chromosomal evolution in higher plants. - Reading, Mass.

STEBBINS, G.L. (1974) - Flowering plants. Evolution above the species level. - Cambridge, Mass.

STURTEVANT, A.H. and BEADLE, G.W. (1940) - An introduction to genetics. - Philadelphia and London.

SWANSON, C.P. (1957) - Cytology and cytogenetics. - Englewood Cliffs.

TIMOFEEFF-RESSOVSKY, N.N., VORONLOV, N.N. & JABLOKOV, A.N. (1975) - Kurzer Grundriss der Evolutionstheorie. - Jena.

TIMOFEEV-RESOVSKIY, N.V., YABLOKOV, A.V. i GLOTOV, N.V. (1973) - Ocherk ucheniya o populyatsii. - Moskva.

TISCHLER, G. (1951) - Allgemeine Pflanzenkaryologie. 2. Hälfte: Kernteilung und Kernverschmelzung. - Berlin-Nikolassee.

TISCHLER, G. und WULFF, H.D. (1953-1963) - Allgemeine Pflanzenkaryologie. Angewandte Pflanzenkaryologie. - Berlin-Nikolassee.

WHITE, M.J.D. (1961) - The chromosomes. 5th edition. - London and New York.

WHITE, M.J.D. (1973) - Animal cytology and evolution. 3rd edition. - London.

WHITEHOUSE, H.L.K. (1965) - Towards an understanding of the mechanism of heredity. - London.

C. **Cytotechnological texts consulted**

ABRAMOVA, L.I., OPEL, L.I. i ORLOVA, I.V. (1971) - Rukovodstvo po tsitologicheskoy teknike. - Leningrad.

BAKER, J.R. (1958) - Principles of biological microtechnique. - London.

BAKER, J.R. (1960) - Cytological technique. 4th edition. - London.

CONN, H.J. (1961) - Biological stains. 7th edition. - Baltimore.

DARLINGTON, C.D. and LA COUR, L.F. (1962) - The handling of chromosomes. 4th edition. - London.

EMIG, W.H. (1959) - Microtechnique. - Colorado Springs.

GALIGAR, A.E. and KOZLOFF, E.N. (1964) - Essentials of practical microtechnique. - Philadelphia.

GEITLER, L. (1940) - Schnellmethoden der Kern- und Chromosomenuntersuchung. - Berlin.

GRAY, P. (1964) - Handbook of basic microtechnique. 3rd edition. - New York.

GRAY, P. (1967) - The use of the microscope. - New York.

GURR, E. (1960) - Encyclopedia of microscopic stains. - Baltimore.

GURR, E. (1967) - Staining: practical and theoretical. - Baltimore.

HASKELL, G. and WILLIS, A.B. (1968) - Primer of chromosome practice. - Edinburgh.

JOHANSEN, D.A. (1940) - Plant microtechnique. 3rd edition. - New York.

LEE, B. (1950) - The microtomist's vade-mecum. 11th edition edited by J.B. Gatenby and H.W. Beams. - London.

LÖVE, Á. and SARKAR, P. (1956) - Cytotechnology. - Winnipeg.

McLEAN, R.C. and IVIMEY-COOK, W.R. (1941) - Plant science formulae. - London.

McLEAN, R.C. and IVIMEY-COOK, W.R. (1952) - Textbook of practical botany. - London, New York, Toronto.

PAZOURKOVÁ, Z. a PAZOUREK, J. (1960) - Rychlé metody botanické mikrotecniky. - Praha.

SASS, J.E. (1958) - Botanical microtechnique. - Ames.

SHARMA, A.K. and SHARMA, A. (1965) - Chromosome techniques. Theory and practice. - London.

SPECTOR, W.S. (ed.) (1956) - Handbook of biological data. - Philadelphia.

D. **Some other texts recommended**

BETHKE, E.G. (1969) - Basic drawing for biology students. - Springfield, Ill.

KNUDSEN, J.W. (1966) - Biological techniques. - New York.

NATHO, G.U.J. (1964) - Herbartechnik. - Wittenberg.

SAVILE, D.B.O. (1962) - Collection and care of botanical specimens. - Ottawa.

SOKAL, R.R. and ROHLF, F.J. (1969) - Biometry. - San Francisco.

STEHLI, G. und BRÜNNER, G. (1968) - Pflanzensammeln - aber richtig. 7. Auflage. - Stuttgart.

VOGELLEHNER, D. (1972) - Botanische Terminologie und Nomenklatur. - Stuttgart.

WOODS, R.S. (1944) - The naturalist's lexicon. - Pasadena.

WOODS, R.S. (1966) - An English-classical dictionary for the use of taxonomists. - Pomona.

164

E. **Other literature cited**

ANDERSON, L.E. and LEMMON, B.E. (1972) - Cytological studies of natural intergeneric hybrids and their parental species in the moss genera *Astomum* and *Weissia*. - Ann. Missouri Bot. Gard. 59: 382-416.

BAJER, A. (1957) - Cine-micrographic studies on mitosis in endosperm. III. The origin of the mitotic spindle. - Exp. Cell Res. 13: 493-502.

BAJER, A. (1959) - Change of length and volume of meiotic chromosomes in living cells. - Hereditas 45: 579-596.

BAJER, A. (1961) - A note on the behaviour of spindle fibers at mitosis. - Chromosoma 12: 64-71.

BARY, A. de (1877) - Vergleichende Anatomie der Vegetationsorgane der Phanerogamen und Farne. - Leipzig.

BARY, A. de (1878) - Über apogame Farne und die Erscheinung der Apogamie im Allgemeinen. - Bot. Zeitschr. 36: 449-487.

BATESON, W. (1902) - Mendel's principles of heredity. - Cambridge.

BATESON, W. and SAUNDERS, E.R. (1902) - Experimental studies in the physiology of heredity. - Rep. Evol. Comm. Roy. Soc. London 2: 1-55; 80-99.

BATTAGLIA, E. (1945) - Sulla terminologia dei processi meiotici. - N.G. Bot. Ital. N.S. 52: 42-57.

BATTAGLIA, E. (1947) - Sulla terminologia dei processi apomictici. - N. G. Bot. Ital., N.S. 54: 674-695.

BEADLE, G.W. (1930) - Genetical and cytological studies of mendelian asynapsis in *Zea mays*. - Cornell Univ. Agr. Exp. Sta. Mem. 129: 1-23.

BEADLE, G.W. and McCLINTOCK, B. (1928) - A genic disturbance of meiosis in *Zea mays*. - Science 68: 433.

BELLING, J. (1926) - The iron acetocarmine method for fixing and staining chromosomes. - Biol. Bull. Marine Biol. Lab. Woods Hole 50: 160-162.

BELLING, J. (1927) - The attachments of chromosomes and the reduction division in flowering plants. - J. Genetics 18: 177-205.

BENNETT, E. (1964) - A rapid modification of the Lautour's technique for grass leaf chromosomes. - Euphytica 13: 44-48.

BENTZER, B., BOTHMER, R. v., ENGSTRAND, L., GUSTAFSSON, M. and SNOGERUP, S. (1971) - Some sources of errors in the determination of arm ratios of chromosomes. - Bot. Not. 124: 65-74.

BLACKBURN, K.B. (1933) - Notes on the chromosomes of the duckweeds (Lemnaceae) introducing the question of chromosome size. - Proc. Univ. Durham Philos. Soc. 9: 84-90.

BLAKESLEE, A.F. (1921) - Types of mutations and their possible significance in evolution. - Amer. Nat. 55: 254-267.

BLAKESLEE, A.F. (1922) - Variations in *Datura* due to changes in chromosome number. - Amer. Nat. 56: 16-31.

BOWER, F.O. (1886) - On apospory in ferns. - J. Linn. Soc., Bot. 21: 360-368.

BOWMAN, W. (1840) - On the minute structure and movements of voluntary muscle. - Phil. Trans. 130: 457-463.

BRIDGES, C.B. (1917) - Deficiency. - Genetics 2: 445-465.

BRIDGES, C.B. (1919) - Duplication. - Anat. Rec. 15: 357-358.

BROWN, R. (1833) - Observations on the organs and mode of fecundation in Orchideae and Asclepiadeae. - Trans. Linn. Soc. London 16: 685-745.

CASPERSSON, T. and ZECH, L. (eds.) (1973) - Chromosome identification technique and applications in biology and medicine. - London.

CHENNAVEERAIAH, M.S. (1960) - Karyomorphologic and cytotaxonomic studies in *Aegilops*. - Acta Horti Gotob. 23: 85-178.

CHIARUGI, A. (1933) - La cariologia nelle sue applicazioni a problemi di botanica. - Atti Soc. Ital. Progr. Sci. 3: 1-38.

CLAUSEN, J., KECK, D.D. and HIESEY, W.M. (1945) - Experimental studies on the nature of species. I. Effects of varied environments on western North American plants. - Carnegie Inst. Wash. Publ. 520: 1-452.

CONGER, A.D. and FAIRCHILD, L.M. (1953) - A quick-freeze method for making smear slides permanent. - Stain Technol. 28: 281-283.

DARLINGTON, C.D. (1926) - Chromosome studies in the Scilleae. - J. Genetics 16: 237-251.

DARLINGTON, C.D. (1929) - Chromosome behaviour and structural hybridity in the Tradescantiae. - J. Genetics 21: 207-286.

DARLINGTON, C.D. (1936) - The external mechanics of the chromosomes. - Proc. Roy. Soc. London, B,121: 264-319.

DARLINGTON, C.D. (1939) - Misdivision and the genetics of the centromere. - J. Genetics 37: 341-364.

DARLINGTON, C.D. and LA COUR, L.F. (1940) - Nucleic acid starvation of chromosomes in *Trillium*. - J. Genetics 40: 185-213.

DARLINGTON, C.D. and MOFFETT, A.A. (1930) - Primary and secondary chromosome balance in *Pyrus*. J. Genetics 22: 129-151.

DELONE, L.N. (1922) - Stravniteljno-kariologicheskoe issledovanie vidov *Muscari* Mill. i *Bellevalia* Lapeyr. - Vestn. Tifliss. Bot. Sada, Ser. II,1: 1-32.

DYER, A.F. (1963) - The use of lacto-propionic orcein in rapid squash methods for chromosome preparation. - Stain Technol. 38: 85-90.

EDMAN, G. (1931) - Apomeiosis und Apomixis bei *Atraphaxis frutescens* C. Koch. - Acta Horti Berg. 11: 13-66.

EIGSTI, O.J. and DUSTIN, P. (1967) - Colchicine in agriculture, medicine, biology and chemistry. - Ames.

ERRERA, L. (1888) - Über Zellformen und Seifenblasen. - Bot. Centralbl. 34: 395-398.

EVANS, A.M. (1969) - Interspecific relationships in the *Polypodium pectinatum-plumula* complex. - Ann. Missouri Bot. Gard. 55: 193-293.

FARMER, J.B. and MOORE, J.E.S. (1905) - On the maiotic phase (reduction-divisions) in animals and plants. - Quart. J. Micr. Sci. 48: 489-557.

FAULKNER, J.S. (1972) - Chromosome studies on *Carex* section *Acutae* in northwest Europe. - Bot. J. Linn. Soc. 65: 271-301.

FERNALD, M.L. (1946) - Technical studies on North American plants. IV. Novelties in our flora. - Rhodora 48: 54-60; 65-81.

FLEMMING, W. (1880) - Beiträge zur Kenntnis der Zelle und ihrer Lebenserscheinungen. II. - Archiv f. mikrosk. Anat. 18: 151-259.

FLEMMING, W. (1882) - Zellsubstanz, Kern und Zelltheilung. - Leipzig.

FOCKE, W.O. (1881) - Die Pflanzenmischlinge. Ein Beitrag zur Biologie der Gewächse. - Berlin.

GATES, R.R. (1911) - Pollen formation in *Oenothera gigas*. - Ann. of Bot. 25: 909-940.

GEITLER, L. (1939) - Die Entstehung der polyploiden Somakerne der Heteropteren durch Chromosomenteilung ohne Kernteilung. - Chromosoma 1: 1-22.

GILMOUR, J.S.L. and GREGOR, J.W. (1939) - Demes: a suggested new terminology. - Nature 144: 333.

GODWARD, M.B.E. (1966) - The chromosomes of the algae. - London.

GOLECHA, P. (1968) - Basic dyes and Feulgen staining. - Nucleus, Suppl.: 215-216.

GREGOIRE, V. (1907) - La formation des gemini hétérotypique dans les végétaux. - Cellule 24: 369-420.

GUSTAFSSON, Å. (1943) - The genesis of the European blackberry flora. - Acta Univ. Lund., N.S. 39,6: 1-200.

GUSTAFSSON, M. (1972) - Distribution and effects of paracentric inversions in populations of *Atriplex longipes*. - Hereditas 71: 173-194.

GUSTAFSSON, M. (1973) - Evolutionary trends in the *Atriplex triangularis* group of Scandinavia. I. Hybrid sterility and chromosomal differentiation. - Bot. Not. 126: 345-392.

HAECKER, V. (1892) - Die heterotypische Kernteilung im Zyklus der generativen Zellen. - Ber. Naturf. Ges. Freib. i.B. 6: 160-193.

HAECKER, V. (1897) - Über weitere Übereinstimmungen zwischen den Fortpflanzungs-vorgängen von Tieren und Pflanzen. - Biol. Zentralbl. 17: 689-705; 721-745.

HASKELL, G. and PATERSON, E.B. (1964) - Quick preparation of corolla chromosomes from flower buds. - Nature 203: 673-674.

HAUNOLD, A. (1968) - Venetian turpentine as an aid in squashing and concomitant production of durable chromosome mounts. - Stain Technol. 43: 153-156.

HEIDENHAIN, M. (1894) - Neue Untersuchungen über die Zentralkörper und ihre Beziehungen zum Kern- und Zellprotoplasma. - Archiv mikrosk. Anat. 43: 423-758.

HEILBRONN, A. und KOSSWIG, C. (1938) - Principia genetica. - J. Unified Sci. (Erkenntnis) 8: 229-255.

HEITZ, E. (1928) - Das Heterochromatin der Moose. I. - Jahrb. wiss. Bot. 69: 762-818.

HEITZ, E. (1931) - Die Ursache der gesetzmässigen Zahl, Lage, Form und Grösse pflanzlicher Nukleolen. - Planta 12: 774-844.

HERIBERT-NILSSON, N. (1915) - Die Spaltungserscheinungen der *Oenothera Lamarckiana*. - Acta Univ. Lund. N.S. II, 12,1: 1-132.

HESLOP-HARRISON, J. and HESLOP-HARRISON, Y. (1970) - Evaluation of pollen viability by enzymatically induced fluorescence; intracellular hydrolysis of fluorescein diacetate. - Stain Technol. 45: 115-120.

HOFMEISTER, W. (1849) - Die Entstehung des Embryo der Phanerogamen. - Leipzig.

HOFMEISTER, W. (1851) - Vergleichende Untersuchungen über die Keimung, Entwicklung und Befruchtung der höheren Kryptogamen und über die Befruchtung der Koniferen. - Leipzig.

HOWARD, A. and PELC, S.R. (1953) - Synthesis of desoxyribonucleic acid in normal and irradiated cells and its relation to chromosome breakage. - Heredity 6, Suppl.: 261-273.

JACKSON, R.C. (1971) - The karyotype in systematics. - Ann. Rep. Ecol. Syst. 2: 327-368.

JANSSENS, F.A. (1909) - Spermatogénèse dans les Bataciens. V. La théorie de la chiasmatypie, nouvelle interprétation des cinèses de maturation. - Cellule 25: 387-411.

JOHANNSEN, W. (1903) - Erblichkeit in Populationen und reinen Linien. - Jena.

JOHANNSEN, W. (1909) - Elemente der exakten Erblichkeitslehre. - Jena.

JUEL, O. (1898) - Parthenogenesis bei *Antennaria alpina* (L.) Br. - Bot. Centralbl. 74: 369-372.

JUEL, O. (1900) - Vergleichende Untersuchungen über typische und parthenogenetische Fortpflanzung bei der Gattung *Antennaria*. - Kgl. Sv. Vet. Akad. Handl. 33,5: 1-59.

KIHARA, H. (1930) - Genomanalyse bei *Triticum* und *Aegilops*. I. - Cytologia 1: 263-270.

KIHARA, H. (1940) - Formation of haploids by means of delayed pollination in *Triticum monöcoccum*. - Bot. Mag. Tokyo 54: 178-185.

KIHARA, H. und ONO, T. (1926) - Chromosomenzahlen und systematische Gruppierung der *Rumex*-Arten. - Z. Zellforsch. mikr. Anat. 4: 475-481.

KIHARA, H. und YAMAMOTO, Y. (1932) - Karyomorphologische Untersuchungen an *Rumex acetosa* L. und *Rumex montanus* Desf. - Cytologia 3: 84-118.

KOLTZOFF, N.K. (1934) - The structure of the chromosomes in the salivary glands of *Drosophila*. - Science 80: 312-313.

KOSTOFF, D. (1939) - Autosyndesis and structural hybridity in F_1 hybrids *Helianthus tuberosus* L. $\times$ *Helianthus annuus* L. and their sequences. - Genetica 21: 285-300.

KURNICK, N.B. and RIS, H. (1948) - A new stain mixture: aceto-orcein fast green. - Stain Technol. 23: 17-18.

LANGLET, O. (1927a) - Beiträge zur Zytologie der Ranunculaceen. - Svensk Bot. Tidskr. 21: 1-17.

LANGLET, O. (1927b) - Zur Kenntnis der polysomatischen Zellkerne im Wurzelmeristem. - Svensk Bot. Tidskr. 21: 397-422.

LANGLET, O. (1946) - A handy field method of fixing root-tips. - Svensk Bot. Tidskr. 40: 425-426.

LANGLET, O. (1971) - Two hundred years genecology. - Taxon 20: 653-722.

LEGENDRE, P. (1972) - The definition of systematic categories in biology. - Taxon 21: 381-406.

LEUCKART, K.G.F.R. (1857) - cf. Rieger und Michaelis, 1958 (arrhenotoky).

LEVAN, A. (1938) - The effect of colchicine on root mitosis in *Allium*. - Hereditas 24: 471-486.

LEVAN, A. (1972) - Cytogenetic effects of hexyl mercury bromide in the *Allium* test. - J. Indian Bot. Soc. 50A: 340-349.

LEVAN, A., FREDGA, K. and SANDBERG, A.A. (1964) - Nomenclature of centromeric position on chromosomes. - Herditas 52: 201-220.

LEWIS, H. (1966) - Speciation in flowering plants. - Science 152: 167-172.

LIN, Y.J. and PADDOCK, E.F. (1973) - Ring-position and frequency of chiasma failure in *Rhoeo spathacea*. - Amer. J. Bot. 60: 1023-1027.

LITARDIÈRE, R. de (1925) - Sur l'existence de figures didiploïdes dans le méristème radiculaire du *Cannabis sativa*. - Cellule 35: 19-26.

LÖVE, Á. (1943a) - Cytogenetic studies on *Rumex* subgenus *Acetosella*. - Hereditas 30: 1-136.

LÖVE, Á. (1943b) - A Y-linked inheritance of asynapsis in *Rumex Acetosa*. - Nature 152: 358-359.

LÖVE, Á. (1952) - Preparatory studies for breeding Icelandic *Poa irrigata*. - Hereditas 38: 11-32.

LÖVE, Á. (1954) - The foundations of cytotaxonomy. - VIIIe Congrès Intern. Bot., Paris, Rapp. et Comm. 9-10: 59-66.

LÖVE, Á. (1964) - The biological species concept and its evolutionary structure. - Taxon 13: 33-45.

LÖVE, Á. (1969) - Sex chromosomes in *Acetosa*. - Chromosomes Today 2: 166-171.

LÖVE, Á. and LÖVE, D. (1949) - The geobotanical significance of polyploidy. I. Polyploidy and latitude. - Portug. Acta Biol. (A), R.B. Goldschmidt Jub. Vol.: 273-352.

LÖVE, Á. and LÖVE, D. (1956) - Chromosomes and taxonomy of eastern North American *Polygonum*. - Canad. J. Bot. 34: 501-521.

LÖVE, Á. et LÖVE, D. (1971) - Polyploïdie et géobotanique. - Nat. Canad. 98: 469-494.

LÖVE, Á. and LÖVE, D. (1974) - Origin and evolution of the arctic and alpine floras. - In: J.D. Ives and R. Barry (eds.): The arctic and alpine environment, London: 571-603.

LÖVE, Á., LÖVE, D. and RAYMOND, M. (1957) - Cytotaxonomy of *Carex* section *Capillares*. - Canad. J. Bot. 35: 715-761.

LÖVE, D. (1944) - Cytogenetic studies on dioecious *Melandrium*. - Bot. Not. 1944: 125-213.

LUNDEGÅRDH, H. (1912) - Fixierung, Färbung und Nomenklatur der Kernstrukturen. Ein Beitrag zur Theorie der zytologischen Metodik. - Archiv f. mikrosk. Anat. 80: 223-273.

McCLINTOCK, B. (1934) - The relation of a particular chromosomal element to the development of the nucleoli in *Zea mays*. - Z. Zellforsch. mikrosk. Anat. 21: 294-328.

McCLUNG, C.E. (1900) - The spermatocyte divisions of the Acrididae. - Kansas Univ. Quart. A,9: 73-100.

McFADDEN, E.S. and SEARS, E.R. (1946) - The origin of *Triticum spelta* and its free-threshing hexaploid relatives. - J. Heredity 37: 81-89; 107-116.

McWILLIAMS, J.R. (1964) - Cytogenetics. - In: Barnard (ed.): Grasses and grasslands, London.

MALHEIROS-GARDÉ, N. and GARDÉ, A. (1950) - Fragmentation as a possible evolutionary process in the genus *Luzula*. - Genet. Iberica 2: 257-262.

MARKS, G.E. (1954) - An acetocarmine glycerol jelly for use in pollen fertility counts. - Stain Technol. 29: 277.

MARKS, G.E. (1973) - A rapid HCl/toluidine blue squash technic for plant chromosomes. - Stain Technol. 48: 229-231.

MARKS, G.E. & SCHWEIZER, D. (1974) - Giemsa banding: Karyotype differences in some species of *Anemone* and in *Hepatica nobilis*. - Chromosoma 44: 405-416.

MATÉRN, B. and SIMAK, M. (1969) - On some statistical problems connected with the identification of chromosomes. - Res. Notes Inst. Forest Mathem. and Statistics, Stockholm, 11: 1-81.

MAZIA, D. (1974) - The cell cycle. - Sci. American 230,1: 55-64.

MELCHERS, G. and LABIB, G. (1970) - Die Bedeutung haploider höherer Pflanzen für Pflanzenphysiologie und Pflanzenzüchtung. - Ber. Dtsch. Bot. Ges. 83: 129-150.

MITTWOCH, U. (1967) - Sex chromosomes. - London and New York.

MONTGOMERY, T.H. (1904) - Some observations and considerations upon the maturation phenomena of the germ cells. - Biol. Bull. Marine Biol. Lab. Woods Hole, 6: 137-158.

MORGAN, T.H. and CATTELL, E. (1912) - Data for the study of sex linked inheritance in *Drosophila*. - J. Exp. Zool. 13: 79-101.

MULLER, H.J. (1925) - Why polyploidy is rarer in animals than in plants. - Amer. Nat. 59: 346-353.

MÜNTZING, A. (1930) - Outlines to a genetic monograph of the genus *Galeopsis*. - Hereditas 13: 185-341.

MÜNTZING, A. (1938) - Note on heteroploid twin plants from eleven genera. - Hereditas 24: 487-491.

MÜNTZING, A. (1939) - Chromosomenaberrationen bei Pflanzen und ihre genetische Wirkung. - Z. ind. Abst. Vererbungsl. 76: 323-350.

MÜNTZING, A. (1944) - Cytological studies of extra fragment chromosomes in rye. I. Iso-fragments produced by misdivision. - Hereditas 30: 231-248.

MÜNTZING, A. (1959) - A new category of chromosomes. - Proc. 10th Int. Genet. Congr., Montreal, 1: 453-467.

MURBECK, S. (1897) - Om vegetativ embryobildning hos flertalet Alchemillor och den förklaring öfver formbeständigheten inom slägtet, som densamma innebär. - Bot. Not. 1897: 273-277.

MURBECK, S. (1901) - Parthenogenetische Embryobildung in der Gattung *Alchemilla*. - Acta Univ. Lund., 36,2, No. 7: 1-45.

NÄGELI, C. v. (1884) - Mechanich-physiologische Theorie der Abstammungslehre. - München und Leipzig.

NAVASHIN, M.S. (1927) - Über die Veränderung von Zahl und Form der Chromosomen infolge der Hybridisation. - Z. Zellforsch. mikrosk. Anat. 6: 195-233.

NAVASHIN, S.G. (1912) - O dimorfizme yader v somaticheskikh kletakh u *Galtonia candicans.* - Izvest. Imp. Akad. Nauk, Ser. VI,4: 373-386.

NAVASHIN, S.G. (1921) - Rezyume vozrazheniy na doklad L.N. Delone. - Zhurn. Russk. Bot. Obshch. 6.

NICOLOFF, H. and DASKALOFF, S. (1964) - A method for making squash preparations permanent. - Izvest. Akad. Nauk. Bulg. 17: 503-505.

NITSCH, J.P. et NITSCH, C. (1970) - Obtention de plantes haploîdes à partir de pollen. - Bull. Soc. Bot. France 117: 339-360.

NORDENSKIÖLD, H. (1951) - Cyto-taxonomical studies in the genus *Luzula*. I. Somatic chromosomes and chromosome numbers. - Hereditas 37: 325-355.

ONO, T. (1935) - Chromosomen und Sexualität von *Rumex acetosa.* - Sci. Rep. Tohoku Imp. Univ. Sendai, IV, Biol. 10: 1-210.

ÖSTERGREN, G. and HENEEN, W.K. (1962) - A squash technique for chromosome morphological studies. - Hereditas 48: 332-341.

OWEN, R. (1849) - cf. Rieger und Michaelis 1958 (parthenogenesis).

PAINTER, T.S. and MULLER, H.J. (1929) - The parallel cytology and genetics of induced translocations and deletions in *Drosophila*. - J. Heredity 20: 287-298.

PANDEY, K.K. (1972) - Origin of genetic variation: Regulation of genetic recombination in the higher organisms - a theory. - Theoret. Appl. Genet. 42: 250-261.

PANDEY, K.K. (1973) - Theory and practice of induced androgenesis. - New Phytol. 72: 1129-1140.

PAZOURKOVÁ, Z. (1964) - Sex chromatin in *Rumex acetosa* L. - Preslia 36: 422-424.

PIJL, L. van der (1969) - Principles of dispersal in higher plants. - Berlin, Heidelberg, New York.

171

POULTON, E.B. (1903) - What is a species? - Proc. Roy. Entom. Soc. London 1903: LXXVI-CXVI.

POWELL, J.B. (1968) - Karyological leaf squashes in grasses, aided by pectinase digestion at 45 C. - Stain Technol. 43: 135-138.

RAJHÁTHY, T. and THOMAS, H. (1972) - Genetic control of chromosome pairing in hexaploid oats. - Nature, New Biol. 239: 217-219.

RANDOLPH, L.F. (1928) - Chromosome numbers in *Zea mays* L. - Cornell Univ. Agric. Exp. Sta. Mem. 117: 1-44.

RENNER, O. (1916) - Zur Terminologie des pflanzlichen Generationswechsels. - Biol. Zentralbl. 36: 337-374.

RIDLEY, H.N. (1930) - The dispersal of plants throughout the world. - Ashford.

ROGERS, J.D. (1973) - Polyploidy in fungi. - Evolution 27: 153-160.

SARKAR, P. and STEBBINS, G.L. (1956) - Morphological evidence concerning the origin of the B-genome in wheat. - Amer. J. Bot. 43: 297-304.

SATÔ, D. (1942) - Karyotype alteration and phylogeny in Liliaceae and allied families. - Jap. J. Bot. 12: 57-161.

SATÔ, D. (1959) - The prokaryotype and phylogeny in plants. - Sci. Papers Coll. Gen. Educ. Univ. Tokyo 9: 303-327.

SATÔ, D. (1962) - Law of karyotype evolution, with special reference to the protokaryotype. - Sci. Papers Coll. Gen. Educ. Univ. Tokyo 12: 173-210.

SAX, H.J. (1960) - Polyploidy in *Enkianthus* (Ericaceae). - J. Arnold Arb. 41: 191-196.

SAX, K. (1959) - The cytogenetics of facultative apomixis in *Malus* species. - J. Arnold Arb. 40: 289-297.

SCHWEIZER, D. (1973) - Differential staining of plant chromosomes with Giemsa. - Chromosoma 40: 307-320.

SEMMENS, C.J. and BHADURI, P.N. (1941) - Staining the nucleolus. - Stain Technol. 16: 119-120.

SERNANDER, R. (1927) - Zur Morphologie und Biologie der Diasporen. - Nov. Acta Roy. Soc. Sci. Upsala, Vol. extra ordin. edit.

SIEBOLD, C.T.E. (1871) - cf. Rieger und Michaelis 1958 (thelytoky).

SINCLAIR, C. and DUNN, D. (1961) - Surface printing of leaves for phylogenetic studies. - Stain Technol. 36: 299-304.

SMITH, J. (1841) - Notice of a plant which produces seeds without any apparent action of pollen. - Trans. Linn. Soc. London 180.

SÖRENSEN, T. (1958) - Sexual chromosome aberrants in triploid apomictic *Taraxaca*. - Bot. Tidsskr. 54: 1-22.

SÖRENSEN, T. and GUDJÓNSSON, G. (1946) - Spontaneous chromosome-aberrants in apomictic *Taraxaca*. Morphological and cytological investigations. - Danske Vid. Selsk. Biol. Skr. IV,2: 1-48.

STEBBINS, G.L. (1947) - Types of polyploids: their classification and significance. - Adv. in Genetics 1: 403-429.

STRASBURGER, E. (1877) - Ueber Befruchtung und Zelltheilung. - Jena Z. Med. Nat. 11: 435-536.

STRASBURGER, E. (1878) - Ueber Polyembryonie. - Jena Z. Med. Nat. 12: 647-670.

STRASBURGER, E. (1884) - Die Controversen der indirekten Zelltheilung. - Archiv mikrosk. Anat. 23: 246-304.

STRASBURGER, E. (1905) - Typische und allotypische Kernteilung. Ergebnisse und Erörterungen. -Jahrb. wiss. Bot. 42: 1-71.

STURTEVANT, A.H. (1926) - A crossover reducer in *Drosophila melanogaster* due to inversion of a section of the third chromosome. - Biol. Zentralbl. 46: 697-702.

TÄCKHOLM, G. (1922) - Zytologische Studien über die Gattung *Rosa*. - Acta Horti Berg. 7: 97-478.

TARKOWSKA, J.A. (1973) - The nature of cytomixis. - Caryologia 25, Suppl.: 151-157.

TISCHLER, G. (1920) - Über die sogenannten "Erbsubstanzen" und ihre Lokalisation in der Pflanzenzelle. - Biol. Zentralbl. 40: 15-28.

TISCHLER, G. (1937) - On some problems of cytotaxonomy and cytoecology. - J. Indian Bot. Soc. 16: 165-169.

TJIO, J.H. and LEVAN, A. (1950) - The use of oxyquinoline in chromosome analysis. - An. Estac. Exp. Aula Dei 2: 21-64.

TRAUB, H.P. (1953a) - Arabinic acid, a new non-precipitating ingredient in combined staining and mounting media. - Euclides 13: 103-114; 149-159.

TRAUB, H.P. (1953b) - Pure arabinates as the chief non-volatile ingredients in combined staining and mounting media. - Euclides 13: 289-298.

VAARAMA, A. (1950) - The dry preservation of fixed plant material. - Stain Technol. 25: 47-50.

VALENTINE, D.H. (1949) - The units of experimental taxonomy. - Acta Biotheoretica 9: 75-88.

VERWORN, M. (1891) - Die physilogische Bedeutung des Zellkernes. -Pflüg. Archiv ges. Physiol. 51.

VICKERY, R.K. (1959) - Barriers to gene exchange within *Mimulus guttatus*. - Evolution 13: 300-310.

VRIES, H. de (1901) - Die Mutationstheorie. 2 Bände. - Leipzig.

WALDEYER, W. (1888) - Über Karyokinese und ihre Beziehung zu den Befruchtungsvorgängen. - Archiv mikrosk. Anat. 32: 1-122.

WALDEYER, W. (1903) - cf. Rieger, Michaelis and Green 1968 (centromere).

WANSCHER, J.H. (1941) - Partial pollen sterility as a somatic character of the peach. - Roy. Vet. and Agric. Coll. Copenhagen, Yearbook 1941: 91-105.

WEBBER, H.J. (1903) - New horticultural and agricultural terms. - Science 18: 501-503.

WEISMANN, A. (1885) - Die Kontinuität des Keimplasmas. - Jena.

WEISMANN, A. (1891) - Amphimixis oder die Vermischung der Individuen. - Jena.

WESTERGAARD, M. (1940) - Studies on cytology and sex determination in polyploid forms of *Melandrium album*. - Dansk Bot. Arkiv 10,3: 1-131.

WESTERGAARD, M. (1958) - The mechanism of sex determination in dioecious flowering plants. - Adv. in Genetics 9: 217-283.

WET, J.M.J. de, and HARLAN, J.R. (1966) - Morphology of the compilospecies *Bothrio-chloa intermedia*. - Amer. J. Bot. 53: 94-98.

WHITE, M.D.J. (1940) - The origin and evolution of multiple sex-chromosome mechanisms. - J. Genetics 40: 303-336.

WHITE, M.D.J. (1945) - Animal cytology and evolution. - Cambridge.

WHITMANN, C.O. (1887) - Germ layers in *Clepsine*. - J. Morphology.

WILLS, A.B. (1962) - The rapid separation of leaf epidermis strips. - Hort. Res. 1: 120-122.

WILSON, E.B. (1896) - The cell in development and heredity. - New York.

WILSON, E.B. (1900) - The cell in development and heredity. 2nd edition. - New York.

WILSON, E.B. (1906) - Studies on chromosomes. III. The sexual difference of the chromosome groups in Hemiptera, with some considerations on the determination and inheritance of sex. - J. Exp. Zool. 3: 1-40.

WILSON, E.B. (1928) - The cell in development and heredity. 3rd edition. - New York.

WINIWARTER, H. (1916) - Recherches sur l'ovogénèse et l'ovaire des Mammifères (Lapin et Homme). - Archives Biol. Paris 17: 33-199.

WINKLER, H. (1908) - Über Parthenogenesis und Apogamie im Pflanzenreich. - Progr. rei. Bot. 2: 293-454.

WINKLER, H. (1916) - Über die experimentelle Erzeugung von Pflanzen mit abweichenden Chromosomenzahlen. - Z. Botanik 8: 417-531.

WINKLER, H. (1920) - Vererbung und Ursache der Parthenogenese im Pflanzen- und Tierreich. - Jena.

WINKLER, H. (1942) - Über den Biontenwechsel und die Abweichungen von seinem normalen Verlauf. - Planta 33: 1-90.

WITSCH, H. v. (1950) - B-Chromosomen als konstante Bestandteile des Chromosomen-satzes von *Alectorolophus*. - Nachr. Akad. Wiss. Göttingen, math.-physik. Kl., biol.-physiol.-chem. Abt. 1950: 21-29.

WRIGHT, S. (1921) - Systems of mating. - Genetics 6: 111-178.

ZIRKLE, C. (1940) - Combined fixing, staining and mounting media. - Stain Technol. 15: 139-153.

174

INDEX

INDEX

abrupt speciation 14, 52
accessory chromosomes 22
acentric fragment 43
acetic stains 150
aceto-carmine-glycerin 98, 101
aceto-carmine-glycerin jelly 100
aceto-gelatin 155
Acetosa 23
    pratensis 21
A-chromosomes 19
acrocentric chromosome 26
adaptation 41, 42
Aegilops
    comosa 24
    squarrosa 57
agamospermy 37, 65ff
agamospory 65ff, 69
agmatoploid, meiosis 61, 62
agmatoploidy 10, 26, 61, 62
Aizoaceae 16, 76
Alchornea ilicifolia 65
alcoholic fixatives 141ff
algae 10, 52, 61, 69
allele 11
Allium 20
    cernuum 19
allogamous 12
allogamy 64
allopatric 52
alloploid 12
allosomes 21ff
amphidiploid 13
amphimixis 66
anaphase 9, 32, 34
Anderson, E. 57
Anderson, L. E. 54
androgenesis 68
aneuploidy 18, 20, 53
anther cultures 68
antheridia 64, 69
antibiotics 43

antipodals 39
apogamety 68
apomictosis 65ff, 69
apomixis 65ff
archegonium 64, 69
archesporium 69
arrhenotoky 65
artificial colors 20, 43
asexual reproduction 65ff
asexual seed formation 66ff
asexual spore formation 69, 70
Asteraceae 66
asynapsis 36, 37
autoalloploid 14, 53
autogamous 12, 60
autogamy 64
autoploid 12, 14, 53
autosegregation 67
autosomes 21
Avena 53
Bajer 8, 95, 97
barbiturates 17
barley 60
de Bary 66, 67
basic numer 12, 18, 45
Bateson 5, 11
Battaglia 68
B-chromosomes 19, 20, 22, 24, 25, 27
Beadle 37, 50
Bellevalia romana 25
Belling 45
Belling's aceto-carmine 141, 150
Bennett 50, 79
Bentzer 97
Bethke 137
Bhaduri 111, 124
binary fission 6
binucleate pollen 35
biocides 43
biological species 1, 42, 57, 59, 73
biological statistics 99